SEVENTH EDITION

BASIC MATHEMATICS FOR ELECTRONICS

ANSWERS TO EVEN-NUMBERED PROBLEMS

Nelson M. Cooke
Late President
Cooke Engineering Company

Herbert F. R. Adams
Former Chief Electronics Instructor
British Columbia Institute of Technology

Peter B. Dell
Late Chief Electronics Instructor
British Columbia Institute of Technology

T. Adair Moore
Former Electronics Instructor, Idaho State University
School of Vocational-Technical Education

GLENCOE

Macmillan/McGraw–Hill

Lake Forest, Illinois Columbus, Ohio Mission Hills, California Peoria, Illinois

Nelson M. Cooke
Late President
Cooke Engineering Company

Herbert F. R. Adams
Former Chief Electronics Instructor
British Columbia Institute of Technology

Peter B. Dell
Late Chief Electronics Instructor
British Columbia Institute of Technology

T. Adair Moore
Former Electronics Instructor, Idaho State University
School of Vocational-Technical Education

1 2 3 4 5 6 7 8 9 MA 99 98 97 96 95 94 93 92 91

Send all inquiries to: Glencoe Division, Macmillan/McGraw-Hill, 936 Eastwind Drive, Westerville, Ohio 43081

ISBN 0-02-800854-5

Printed in the United States of America.

CONTENTS

PREFACE			v
PRETESTS			vi
Chapter	2	Algebra--General Numbers	1
	3	Algebra--Addition and Subtraction	1
	4	Algebra--Multiplication and Division	2
	5	Equations	5
	6	Powers of 10	7
	7	Units and Dimensions	9
	8	Ohm's Law--Series Circuits	11
	9	Resistance--Wire Sizes	15
	10	Special Products and Factoring	16
	11	Algebraic Fractions	21
	12	Fractional Equations	27
	13	Ohm's Law--Parallel Circuits	39
	14	Meter Circuits	42
	15	Divider Circuits and Wheatstone Bridge	43
	16	Graphs	45
	17	Simultaneous Equations	46
	18	Determinants	56
	19	Batteries	61
	20	Exponents and Radicals	63
	21	Quadratic Equations	67
	22	Network Simplification	74
	23	Angles	82
	24	Trigonometric Functions	84
	25	Trigonometric Values	86
	26	Solution of Right Triangles	87
	27	Trigonometric Identities and Equations	89
	28	Elementary Plane Vectors	93
	29	Periodic Functions	95
	30	Alternating Currents--Fundamental Ideas	96
	31	Phasor Algebra	98
	32	Alternating Currents--Series Circuits	100
	33	Alternating Currents--Parallel Circuits	106
	34	Logarithms	118
	35	Applications of Logarithms	121
	36	Number Systems for Computers	129
	37	Boolean Algebra	138
	38	Karnaugh Maps	144

PREFACE

The question has been asked whether the problems in *Basic Mathematics for Electronics*, Seventh Edition, by Cooke, Adams, Dell, and Moore are to be solved by means of computers and/or calculators. The answer is that the problems may be computed by any means: paper and pencil, tables, slide rules, abacus, calculator, computer--or mentally, for that matter. The point is that the problems should be solved according to the needs of the students, the requirements and objectives of the instructor, the traditions and requirements of industry, the complexity of individual problems, and the availability of the various computational aids.

All the answers in this answer key were computed on hand-held calculators, and they have been checked against those previously calculated by slide rule and interpolated five-place tables.

Certainly the solutions may differ with the calculating mode. In the past a slide rule gave sufficient accuracy when the solution was to be converted into a real preferred value component, ± 10%. Five-place tables can deliver an answer which is in general agreement to the first four significant figures. One brand of calculator may guarantee a closer or more accurate figure than another brand. Nothing is to be gained from an argument between an enthusiast for reverse Polish notation (RPN) and an equally enthusiastic advocate of algebraic notation, whose calculations differ in the eighth or tenth significant figure. The working value of real meter readings, real components, and real measurable performance is what matters.

The authors will welcome comments from users of this manual. They are especially interested in suggestions for its improvement and in corrections of any errors that may have escaped them.

PRETESTS

The phenomenal growth in the use of electronic calculators and computers appears, at times, to have eliminated the need for technicians to perform arithmetical calculations unaided by the products of the electronics industry. However, there are times when even workers in the field will find that their calculator batteries have run down or their calculators and computers themselves are not at hand.

We therefore suggest that these pretests be performed without the aid of calculators and/or computers. We believe that all technicians should demonstrate a grasp of the fundamentals of arithmetic, if only so they can, in the future, compute reasonable approximations to check the more complete calculations performed electronically.

Basic Mathematics for Electronics is offered in the expectation that its students have demonstrable ability to perform the basic arithmetical operations. We suggest that the students take these tests at the outset of their advanced education for several reasons:

1. To enable them to demonstrate their preparedness to benefit from *Basic Mathematics for Electronics*.

2. To indicate to them areas in which early reinforcement of previous schooling will contribute to greater success in their new studies.

3. To help the instructor assess the abilities of new students.

4. To indicate to the instructor where to begin teaching and how to subdivide classes and work load.

Students should be instructed to copy the problems into their workbooks *correctly*. One of the most serious continuing sources of error is the incorrect copying of data. Encourage your students to develop the habit of copying data and immediately checking them for correctness.

Pretest 1. Whole Numbers: Addition and Subtraction

Add:

1. 120 + 47 + 5162

2. 18 900 + 2587 + 143

3.
```
  2 735
     54
 19 647
```

4.
```
 2 073 162
 5 108 009
     5 017
   109 463
```

5.
```
 18 441 036
  1 000 055
      2 189
 12 947 086
  5 431 255
```

Subtract:

6. 247 from 1862

7. 2755 from 18 032

8.
```
 36 091
  8 463
```

9.
```
 41 087
 23 918
```

10.
```
 10 008
  6 939
```

11. The sum of 6934 and 2175 from the sum of 9286 and 3004.

12. The sum of 1962 and 2699 from the sum of 1847 and 3255.

13. A wholesaler of electronic parts sold 12 850 resistors to a vocational college, 165 to a broadcast station, and 927 to an amateur radio club. His stock check then showed another 18 643 on hand. How many resistors did he have at first?

14. How much greater than 108 697 is 203 108?

15. In a dc circuit the sum of the voltage drops must equal the sum of the applied electromotive forces (emf). How much emf in the form of battery voltage must be added to a circuit which now contains emf's totalling 24 V to meet the requirements of the following list of voltage drops: 5.6 V, 1.2 V, 3.8 V, and 8.5 V?

Pretest 2. Whole Numbers: Multiplication and Division

Multiply:

1. 24 by 25

2. 180 by 15

3. $\begin{array}{r} 1807 \\ \underline{16} \end{array}$ 4. $\begin{array}{r} 5086 \\ \underline{227} \end{array}$ 5. $\begin{array}{r} 4283 \\ \underline{9537} \end{array}$

Divide:

6. 1024 by 8

7. 3655 by 5

8. 7987 by 49

9. $627 \overline{\smash{)}3\,094\,872}$

10. $639 \overline{\smash{)}1\,730\,412}$

11. $100 \overline{\smash{)}100\,000\,000}$

12. Divide the product of 872 and 35 by 5.

13. Divide the product of 186 and 44 by the product of 31 and 12.

14. Divide the sum of the products of 17 and 20 and 7 and 60 by the product of 19 and 2.

15. The impedance of an ac circuit, in ohms, may be found by dividing the applied emf, in volts, by the current, in amperes. What is the impedance of a circuit which limits the current to 3 A when the applied emf is 24 V?

Pretest 3. Fractions: Addition and Subtraction

Add:

1. $\frac{3}{32} + \frac{5}{32}$

2. $\frac{5}{16} + \frac{2}{16}$

3. $\frac{5}{8} + \frac{1}{16}$

4. $\frac{1}{4} + \frac{1}{3}$

5. $\frac{15}{64} + \frac{5}{32} + \frac{3}{8}$

6. $9\ \frac{5}{100} + \frac{7}{10} + 1\ \frac{650}{1000}$

Subtract:

7. $\frac{56}{64} - \frac{32}{64}$

8. $\frac{27}{32} - \frac{5}{32}$

9. $\frac{5}{8} - \frac{3}{16}$

10. $\frac{1}{2} - \frac{3}{8}$

11. $\frac{45}{100} - \frac{130}{1000}$

12. $5\ \frac{7}{100} - 2\ \frac{83}{1000}$

13. 2 ft $5\ \frac{3}{16}$ in - 1 ft $7\ \frac{15}{32}$ in

14. 2 m 80 cm 12 mm - 1 m 90 cm 27 mm

15. 4652 mm - 1536 mm

Pretest 4. Fractions: Multiplication and Division

Multiply:

1. $\frac{1}{4}$ by $\frac{2}{3}$

2. $\frac{3}{4} \times \frac{5}{9}$

3. $\frac{9}{10} \times \frac{7}{10}$

4. $\frac{1}{2} \times \frac{7}{64}$

5. $3\frac{3}{10} \times 2\frac{75}{100}$

Divide:

6. $\frac{7}{8}$ by $\frac{3}{16}$

7. $\frac{5}{8} \div \frac{5}{32}$

8. $\frac{35}{75} \div \frac{28}{100}$

9. $4\frac{2}{3} \div 2\frac{5}{12}$

10. $5\frac{3}{10} \div 4\frac{24}{100}$

Pretest 5. Decimals: Addition and Subtraction

Add:

1. 24.37 + 18.95

2. 86.01 + 222.63 + 15.88

3.
```
  0.005 27
  1.906 85
  ________
```

4.
```
  87.016
   4.942
  ______
```

5.
```
   27.136
  192.030 7
   91.100 9
  _________
```

Subtract:

6. 0.003 92 from 1.091 76

7. 2.105 from 6.003

8. Subtract 1.270 from the sum of 0.9877 and 0.3926.

9. Subtract the sum of 2.741 53 and 54.830 52 from 61.200 59.

10. Subtract the sum of 4.271 6 and 2.008 7 from the sum of 3.110 9 and 5.019 7.

Pretest 6. Conversion: Common Fractions to Decimals

Write the decimal equivalent of:

1. $\frac{3}{8}$

2. $\frac{7}{8}$

3. $\frac{4}{25}$

4. $\frac{5}{9}$

5. $5\ \frac{287}{1000}$

Pretest 7. Conversion: Decimals to Common Fractions

Write the common-fraction equivalent with the given denominator (to the nearest whole number numerator):

1.	0.4	5
2.	0.375	8
3.	0.4256	7
4.	0.531 25	32
5.	0.8906	64

Pretest 8. Percentage

1. What is 15% of $2.00?
2. What is 80% of $41.24?
3. $31.67 is what percent of $47.50?
4. 177 is what percent of 885?
5. What percent of 1 m is 1 cm?
6. An article marked $287.88 is sold at a trade discount of 30%. The sale price is subject to a local tax of 7.18%. What is the total cost?
7. Successive discounts are each based on the "new" price. What is the net cost of an article marked "$298.88 less 25% and 10%"?
8. Successive markups are each based on the "new" price. What is the total cost of a $25.00 article purchased in country A, subject to 30% duty and 12% sales tax on the duty-paid value when imported into country B if the rate of exchange makes $1.00 of country B worth $0.88 in country A?
9. A resistor is rated "470 Ω ± 10%." Within what range of values may it be considered acceptable?
10. A resistor color-coded 5.6 kΩ has a measured value of 5.24 kΩ. What is the percentage error from its "apparent" value?

Pretest 9. Powers and Roots

Evaluate:

1. 16^2
2. 5.5^2
3. $\left(\frac{5}{8}\right)^2$
4. 2^3
5. 2^6
6. $\sqrt{144}$
7. $\sqrt{3401.22}$
8. $\sqrt{0.746\ 669}$
9. $\sqrt{160\ 000}$
10. $\sqrt[3]{64}$

Pretest 10. Positive and Negative Numbers

Add:

1. 53 and 41
2. -67 and -14
3. 122 and -95
4. 122 and -419
5. -32.41 and 68.57

Subtract:

6. 37 from 118
7. 118 from 37
8. 42 from -77
9. -146 from 188
10. -15.88 from -53.47

Multiply:

11. 2.5 by 7.3
12. 15.2 by 16.7
13. 1.9 by -12.7
14. -3.97 by -8.22
15. -5.13 by -8.67

Divide:

16. 24 by 6
17. -42 by 6
18. 136 by -4
19. -56 by -7
20. -38.08 by -2.8

Pretest 11. Significant Figures

To how many significant figures are the following numbers written?

1. 3.456
2. 0.001 234
3. 86 240
4. 123 400.000

Round off the following numbers to the indicated number of significant figures:

5. 45.782 (3)
6. 1.011 625 (6)
7. 288.985 (5)

Write the following numbers as numbers between one and 10 times the appropriate power of 10:

8. 28 000
9. 215 416.920 6
10. 0.000 028 65

ANSWERS TO PRETESTS

Pretest 1

1. 5329
2. 21 630
3. 22 436
4. 7 295 651
5. 37 821 621
6. 1615
7. 15 277
8. 27 628
9. 17 169
10. 3 069
11. 3181
12. 441
13. 32 585
14. 94 411
15. 4.9 V

Pretest 2

1. 600
2. 2700
3. 28 912
4. 1 154 522
5. 40 846 971
6. 128
7. 731
8. 163
9. 4936
10. 2708
11. 1 000 000
12. 6104
13. 22
14. 20
15. 8 Ω

Pretest 3

1. $\frac{8}{32}\left(= \frac{1}{4}\right)$
2. $\frac{7}{16}$
3. $\frac{11}{16}$
4. $\frac{7}{12}$
5. $\frac{49}{64}$
6. $11\ \frac{400}{1000}\ \left(= 11\ \frac{2}{5}\right)$
7. $\frac{3}{8}$
8. $\frac{11}{16}$
9. $\frac{7}{16}$
10. $\frac{1}{8}$
11. $\frac{8}{25}$
12. $2\ \frac{987}{1000}$
13. $9\ \frac{23}{32}$ in
14. 885 mm
15. 3116 mm

Pretest 4

1. $\frac{1}{6}$
2. $\frac{5}{12}$
3. $\frac{63}{100}$
4. $\frac{7}{128}$
5. $9\ \frac{75}{1000}$
6. $\frac{14}{3}$
7. 4
8. $1\ \frac{2}{3}$
9. $1\ \frac{27}{29}$
10. $1\ \frac{1}{4}$

Pretest 5

1. 43.32
2. 324.52
3. 1.912 12
4. 91.958
5. 310.267 6
6. 1.087 84
7. 3.898
8. 0.1103
9. 3.628 54
10. 1.8503

Pretest 6

1. 0.375
2. 0.875
3. 0.16
4. 0.5556
5. 5.287

Pretest 7

1. $\frac{2}{5}$
2. $\frac{3}{8}$
3. $\frac{3}{7}$
4. $\frac{17}{32}$
5. $\frac{57}{64}$

Pretest 8

1. \$0.30
2. \$32.99
3. 66.7%
4. 20%
5. 1%
6. \$215.98
7. \$201.74
8. \$41.36
9. 423 to 517 Ω
10. -6.43%

Pretest 9

1. 256
2. 30.25
3. $\frac{25}{64}$ or (0.390 625)
4. 8
5. 64
6. ±12
7. ±58.32
8. ±0.8641
9. ±400
10. 4

Pretest 10

1. 94
2. -81
3. 27
4. -297
5. 36.16
6. 81
7. -81
8. -119
9. 334
10. -37.59
11. 18.25
12. 253.84
13. -24.13
14. 32.63
15. 44.48
16. 4
17. -7
18. -34
19. 8
20. 13.6

Pretest 11

1. 4
2. 4
3. 5
4. 9
5. 45.8
6. 1.011 63
7. 288.98
8. 2.80×10^4
9. 2.15×10^5
10. 2.87×10^{-5}

PROBLEMS 2-1

2. (a) 35 A
(b) 64 Ω
(c) 1760 V

4. (a) 17 cents
(b) 17 *p* cents

6. 2 R Ω and 12 R Ω

8. (a) 15
(b) 14
(c) 2.4
(d) $\frac{1}{6}$
(e) 3

10. 750 - *R* Ω

12. (a) 3.75 pF
(b) 114 pF

14. 0.000 827 s

16. (a) 0.306 km
(b) 2.47 m

PROBLEMS 2-2

2. (a) 3300
(b) 99 000
(c) 6708
(d) 3530
(e) 10 225

4. 1(b), (d), (f);
2(c), (d), (e);
3(d), (e), (f), (g), (h), (j)

6. 8.73 mH

8. (a) 625 W
(b) 450 W

10. (a) multiplied by a factor of 4
(b) divided by a factor of 4
(c) divided by a factor of 2
(d) multiplied by a factor of 2

PROBLEMS 3-1

2. 18

4. -65

6. -407

8. -0.0045

10. -118.21

12. $-4\frac{1}{8}$

14. $2\frac{5}{8}$

PROBLEMS 3-2

2. 121

4. -377

6. -0.078 12

8. $4\frac{1}{16}$

10. $-1\frac{1}{8}$

12. (a) 180°
(b) 71°
(c) 36°

14. 685 V

PROBLEMS 3-3

2. $-12i^2r$
4. $55.9IR$
6. $30R - jX$
8. $16\Omega - 14\omega$
10. $18X$, $15R$, $-9L$, Q
12. $31IX - 5IZ + 10IR$
14. $2.15vi + 2.21\frac{v^2}{r} - 1.88i^2r$
16. $32IR - IZ$
18. $12.7\omega L + 8.2X_c - 18.8Z$
20. $-22.8X_c - \frac{2.78}{\omega c}$
22. $8iR - 2V_g$
24. $14.5a - 11.4b + 5.4c$

PROBLEMS 3-4

2. $9\lambda - 2\theta - 6$
4. $5I^2R - 15VI + 5$
6. $A - 12B + 2C$
8. $3\theta - 3\phi + 3\omega$
10. $16.72R - 7.2Z - 7.2IX + 3X_c - 27$

PROBLEMS 3-5

2. (a) $5\omega - (-6x + 3y - 4z)$
   (b) $2X_L + 7X_c - (5R_1 - 4Z - 3R_2)$
   (c) $2I_1 - (-7I_2 + 4I_3 + 7I_4)$
   (d) $0.004IR - (2V - 1.6\frac{P}{I} + 0.006IZ)$
   (e) $25\frac{V^2}{R} + W - (4I^2Z + 12VI + 18I^2R)$
4. $526 - i$ mA
6. $X_L - \frac{1}{2\pi f C}$
8. $Q - (X - 2\pi f L)$
10. $\beta - 19.6$
12. $R^2 = Z^2 - X^2$

PROBLEMS 4-1

2. -30
4. 94.772
6. $\frac{7}{16}$
8. 6.75
10. $qrst$
12. $\theta^2\phi^2\lambda^2$
14. $-\frac{1}{abcd}$

PROBLEMS 4-2

2. $-b^8$
4. L^3Y^5
6. $-240\phi^2$
8. $25\mu^2$
10. $-26b^{a+x+y}$
12. $-125\lambda^6$

14. $-\frac{a^4b^2}{6}$

16. $-4a^3b^5cd^2e$

18. $\frac{\theta^4\mu\phi\omega}{30}$

20. $3^r p^{qr}$

PROBLEMS 4-3

2. $6a^2 + 9a$

4. $3.2cd^2 - jd^3$

6. $-10\alpha^4 + 15\alpha^3 - 20\alpha^2$

8. $3x^3y^2 - x^3y^3 - 2x^2y^3$

10. $-30\omega^3L_1^2L_2 + 60\omega^3L_1L_2M - 15\omega^3L_1L_2^2$

12. $-2a^2b^3 - a^2b^4 - ab^2$

14. $-15\theta k^2\mu^2\omega - 9\eta\theta k\omega + 6\eta^2\theta\mu\omega$

16. $-12a^2r^3s + 6ar^2s^2 - 18a^2rs^3$

18. $2\alpha\mu$

20. $-4XY^2 + 13a^2X$

22. $2\gamma^3\beta^2 - \gamma^2\beta^3 - 5\gamma^4\beta + 2.5\gamma\beta^4$

24. $-8\theta^3 + 21\theta^2\phi - 9\theta\phi^2$

26. $3\lambda^3\mu^2 - 2.5\lambda^2\mu^3 + 6\lambda^2\mu - 2\lambda\mu^2$

28. $\frac{5\gamma^3\lambda}{3} - \beta\lambda^2\gamma^2 + 3\beta^3\lambda + \frac{6\gamma\beta^2\lambda}{5} - \gamma^2\beta^2\theta$

30. $2.16P^2R_1 + 4.8P^2R_2 - 1.576P^2R_3$

PROBLEMS 4-4

1. $\alpha^2 + 2\alpha + 1$

3. $\alpha^2 - 2\alpha + 1$

5. $\beta^2 - 9$

2. $\alpha^2 - 1$

4. $\beta^2 + 6\beta + 9$

6. $\beta^2 - 6\beta + 9$

7. $x^2 + 7x + 12$

9. $p^2 - 8p - 48$

8. $L_1^2 - 8L_1 + 15$

10. $a^2 - 25$

12. $15X^2 + 8XZ + Z^2$

14. $acx^2 + bdx^2 + adx^2 + bcx^2$

16. $45V^2I^2 - 33VI^3R + 6I^4R^3$

18. $3\psi^2 + 3.625\psi\phi + 0.875\phi^2$

20. $\frac{m^2}{4} - \frac{7mq}{72} - \frac{5q^2}{12}$

22. $6L_1^3 + 16L_1^2 - 13L_1 - 10$

24. $x^3 + 3x^2y + 3xy^2 + y^3$

26. $p^3 - 3p^2q + 3pq^2 - q^3$

28. $I^3R^3 - I^2R^2P - IRP^2 + P^3$

30. $a^2 + 2a + 1$

32. $x^2 - 2xy + y^2$

34. $4\theta^2\phi^2 + 4\theta\phi\psi + 4\theta\phi + 2\psi + \psi^2 + 1$

36. $11\alpha^2 - 22\alpha - 32$

38. $30X^2 - 64XY - 2XZ - 24Y^2 + 58YZ + 2Z^2$

40. $74\omega^2\theta^2 - 44\omega\lambda\theta^2 + 34\omega\theta^3 - 46\lambda^2\theta^2 + 22\lambda\theta^3 - \theta^4$

PROBLEMS 4-5

2. -4
4. -0.03
6. $\frac{3}{4}$
8. $\frac{1}{2\pi f c}$
10. $\frac{\omega L}{Q}$
12. -8
14. 81

PROBLEMS 4-6

2. $8a^3y^2z^2$
4. $4\omega^2L^2$
6. $-100a^7b^4c^{10}d^3$
8. $-5\ell^2m^4p^2$
10. $\frac{5y^2}{6xz^2}$
12. $\frac{4\phi}{\theta^2\alpha^2}$
14. $-90V^3E^4(IR)^5$
16. $\frac{s^4u^4}{3rt}$
18. $\frac{7x^5y}{6\alpha^2}$
20. $-\frac{0.016I^6R^4}{Z^2}$

PROBLEMS 4-7

2. $3\theta - 2\phi$
4. $5x^4 - 4x^2 + 2$
6. $3\Delta^5 - 5\Delta^4 - 7\Delta$
8. $\frac{90\alpha}{\gamma} + \frac{80\gamma}{\beta} - \frac{40}{\alpha\beta\gamma}$
10. $\frac{1}{2} - \frac{5I^2R}{12} - \frac{2}{5I^4R^2} - \frac{1}{I^6R^3}$
12. $X^2 + x^2$
14. $-(\alpha - \beta) = \beta - \alpha$
16. $6I^7(R + r)^3(R - r)^3 - 5I^3(R + r)(R - r) - 3I$
18. $-4\alpha^2 - 8\alpha - 1$
20. $6V + 4V^3(R + R_1)(r + r_1) - 12V^5(R + R_1)^2(r + r_1)^2$

PROBLEMS 4-8

2. $4m + 6$
4. $4Q + 5$
6. $5Q - 2E$
8. $2\phi^2 + 2\omega^2$
10. $5k^2 - 2k\lambda - 6\lambda^2$
12. $E^2 + Ee + e^2$
14. $L^2 + I^2X^2$
16. $\theta^4 - \theta^3\phi + \theta^2\phi^2 - \theta\phi^3 + \phi^4$
18. $\alpha^2 + \beta^2$
20. $P^3 - 2P^2Q - 2PQ^2 + Q^3$

22.

$$
\begin{array}{r|l}
 & 2m^2 + m - 1 \\
\hline
m^2 - m + 3 & 2m^4 - m^3 + 4m^2 + 7m + 1 \\
 & 2m^4 - 2m^3 + 6m^2 \\
\cline{2-2}
 & \quad m^3 - 2m^2 + 7m + 1 \\
 & \quad m^3 - m^2 + 3m \\
\cline{2-2}
 & \qquad -m^2 + 4m + 1 \\
 & \qquad -m^2 + m - 3 \\
\cline{2-2}
 & \qquad\qquad 3m + 4 \quad \text{remainder}
\end{array}
$$

$$\text{Answer} = 2m^2 + m - 1 + \frac{3m + 4}{m^2 - m + 3}$$

24. $\frac{\theta^2}{4} - 3\theta\phi + 9\phi^2$

26.

$$
\begin{array}{r|l}
 & n^2 - \frac{6}{5}n - \frac{27}{25} \\
\hline
n - \frac{3}{5} & n^3 - \frac{9}{5}n^2 - \frac{9}{25}n - \frac{27}{125} \\
 & n^3 - \frac{3}{5}n^2 \\
\cline{2-2}
 & \quad - \frac{6}{5}n^2 - \frac{9}{25}n - \frac{27}{125} \\
 & \quad - \frac{6}{5}n^2 + \frac{18}{25}n \\
\cline{2-2}
 & \qquad - \frac{27}{25}n - \frac{27}{125} \\
 & \qquad - \frac{27}{25}n + \frac{81}{125} \\
\cline{2-2}
 & \qquad\qquad - \frac{108}{125}
\end{array}
$$

$$\text{Answer} = n^2 - \frac{6}{5}n - \frac{27}{25} - \frac{\frac{108}{125}}{n - \frac{3}{5}}$$

28. $\frac{K^2}{9} - \frac{K}{6} + \frac{1}{16}$

30. $R_1^6 - \frac{R_1^5 V}{I} + R_1^4\left(\frac{V}{I}\right)^2 - R_1^3\left(\frac{V}{I}\right)^3 + R_1^2\left(\frac{V}{I}\right)^4 - R_1\left(\frac{V}{I}\right)^5 + \left(\frac{V}{I}\right)^6$

PROBLEMS 5-1

2. $\theta = 4$ **4.** $L = 3$ **6.** $\lambda = 3$ **8.** $M = 1$ **10.** $Q = 3$

12. $\lambda = 7$ **14.** $L = 7$ **16.** $R_1 = 0.50$ **18.** $\phi = 4$ **20.** $Z = -5$

PROBLEMS 5-2

2. $R + 68.8\ \Omega$

4. $f = \dfrac{n}{n + 8}$

6. $12\ell - s$ months

8. 35

10. $h = \dfrac{v}{\ell\omega}$ cm

12. $P = I^2R$

14. Meter + Scope = \$574.00, Meter = Scope - 356

$\therefore s - 356 + s = \$574.00$ Scope = \$465.00

Meter cost = 574 - 465 = \$109.00

16. Smallest angle $(a) = \dfrac{b}{2}$ $\quad \therefore b = 2a$

Largest angle $(c) = a + 52^\circ$

$a + b + c = 180^\circ,\ a + 2a + a + 52^\circ = 180^\circ \therefore a = 32^\circ$

$b = 2a = 64^\circ,\ c = a + 52^\circ = 84^\circ$

Answer $a = 32^\circ,\ b = 64^\circ,\ c = 84^\circ$

18. $15 + 16 = 31$

20. $VI = \dfrac{V^2}{R}$

PROBLEMS 5-3

2. $I = \dfrac{V}{Z},\ Z = \dfrac{V}{I}$

4. $R = \dfrac{P}{I^2},\ I^2 = \dfrac{P}{R}$

6. $R_1 = R_t - R_2 - R_3 \qquad R_2 = R_t - R_1 - R_3 \qquad R_3 = R_t - R_1 - R_2$

8. $r = \dfrac{c}{2\pi},\ \pi = \dfrac{c}{2r}$

10. $C = \dfrac{1}{2\pi f X_C},\ f = \dfrac{1}{2\pi C X_C}$

12. $r = \dfrac{s}{2\pi h},\ h = \dfrac{s}{2\pi r}$

14. $N_p = \dfrac{V_p N_s}{V_s},\ V_s = \dfrac{V_p N_s}{N_p},\ V_p = \dfrac{V_s N_p}{N_s}$

16. $h = \dfrac{T - 2A}{P},\ T = 2A + ph$

18. $F = LHi,\ L = \dfrac{F}{Hi},\ i = \dfrac{F}{LH}$

20. $g_m = \dfrac{\mu}{r_p}$

22. $g = \dfrac{V^2}{2h},\ V^2 = 2gh$

24. $\omega = 2\pi n$

26. $\ell = \dfrac{8\mu\omega}{B^2A},\ \omega = \dfrac{B^2A\ell}{8\mu},\ A = \dfrac{8\omega\mu}{B^2\ell}$

28. $m = \dfrac{F}{4\pi^2n^2r},\ F = 4\pi^2n^2mr$

30. $T = \dfrac{tC}{C - F},\ F = \dfrac{TC - tC}{T}$

32. $R = PFX,\ PF = \dfrac{R}{X}$

34. $k = \dfrac{M}{\sqrt{L_1L_2}} \Rightarrow \dfrac{M\sqrt{L_1L_2}}{L_1L_2}$

36. $Tg = \dfrac{vI}{2kF}$

38. $P_{no} = \dfrac{P_{so}P_{ni}}{P_{si}}$

40. $V_{pt} = V_b W - V_B$

42. $b = \dfrac{4av - h}{2}$

44. $G_o = G + \dfrac{g_m}{1 + n}$

46. $R = \dfrac{V^2}{P} = \dfrac{117^2}{114}$ $\therefore R = 120\ \Omega$

48. 2 poles (1 pair)

50. 177 MHz

PROBLEMS 5-4

2. $\frac{2}{3}$ **4.** $\frac{20}{1}$ **6.** 1.25:10, 125:1000, etc. **8.** $\frac{1}{4}$ **10.** $\frac{1}{5}$

PROBLEMS 5-5

2. 36 **4.** 16 **6.** $q = 5$ **8.** $d = 0.8$ **10.** $z = 8$

PROBLEMS 5-6

2. $C \propto W,\ C = kW$

4. $X_L \propto fL,\ X_L = kfL$

6. $R \propto \dfrac{L}{A},\ R = \dfrac{kL}{A}$

8. $V \propto r^3,\ V = kr^3$

10. $\dfrac{V_1}{V_2} = k\left(\dfrac{\ell_1}{\ell_2}\right)^3$

12. $I = kV,\ k = \dfrac{I}{V} = \dfrac{1.2}{60},\ I = \dfrac{1.2}{60} \times 85 = 1.7\text{ A}$

14. $R = \dfrac{k\ell}{d^2},\ k = \dfrac{5.21 \times 2.05^2}{1000} = 0.0219$

$$R = \frac{5.21 \times 2.05^2}{1000} \times \frac{500}{2.588^2} = 1.63\ \Omega$$

PROBLEMS 6-1

2. 3 **4.** 6 **6.** 1 **8.** 3 **10.** 7

PROBLEMS 6-2

2. 1.36×10

4. 9.63×10^{-2}

6. 8.74×10^{6}

8. 5.92×10^{4}

10. 8.60×10

12. 4.00×10^{4}

14. 3.14×10^{-2}

16. 1.25×10^{-6}

18. 8.15×10^{6}

20. 5.56×10^{-8}

21. (2) 13.6

(4) 96.3×10^{-3}

(6) 8.74×10^{6}

(8) 59.2×10^{3}

(10) 86.0

(12) 40×10^{3}

(14) 31.4×10^{-3}

(16) 1.25×10^{-6}

(18) 8.15×10^{6}

(20) 55.6×10^{-9}

PROBLEMS 6-3

2. 1.00×10^{2}

4. 8.00×10^{-1}

6. 1.31×10^{-3}

8. 1.45×10^{-9}

10. 1.84×10^{11}

12. $9.42 \times 10\ \Omega$

14. $5.65 \times 10^{2}\ \Omega$

PROBLEMS 6-4

2. 1.00×10^{0}

4. 1.68×10^{3}

6. 2.27×10^{-3}

8. 7.93×10^{-2}

10. 7.96×10^{0}

12. 2.81×10

14. $2.27 \times 10^{2}\ \Omega$

PROBLEMS 6-5

2. 10^{-12}

4. 1.6×10^{-7}

6. 2.7×10^{-5}

8. 1.6×10^{-1}

10. 1.8×10^{-3}

12. 6.0×10^{3}

14. 3.23×10^{7}

16. 3.75×10^{6} Hz

18. 4.69×10^{5} Hz

20. 9.51 Hz

PROBLEMS 6-6

2. (a) 25 003 400

(b) 2.50×10^{7}

4. (a) 2 000.4

(b) 2.00×10^{3}

6. 50.2 nF

or 50 200 pF

PROBLEMS 7-1

2. (a) 6.85×10^3 mA

(b) 6.85×10^6 μA

4. (a) 1.25×10^5 μA

(b) 1.25×10^{-1} A

6. (a) 5.00×10^{-5} F

(b) 5.00×10^7 pF

8. (a) 1.65×10^{-2} H

(b) 1.65×10^4 μH

10. (a) 2.53×10^4 ms

(b) 2.53×10^7 μs

12. (a) 4.70×10^4 Ω

(b) 4.70×10^{-2} MΩ

(c) 2.13×10^{-5} S

14. (a) 2.20×10^{-1} mH

(b) 2.20×10^{-4} H

16. (a) 8.00×10^{-3} ms

(b) 8.00×10^{-6} s

(c) 8.00×10^3 ns

18. (a) 3.25×10^5 V

(b) 3.25×10^{-1} MV

20. (a) 3.70×10^3 W-h

(b) 3.70×10^6 mW-h

22. (a) 5.06×10^5 kHz

(b) 5.06×10^8 Hz

24. (a) 1.50×10^6 μs

(b) 1.50×10^0 s

(c) 1.50×10^9 ns

26. (a) 5.00×10^{-3} S

(b) 2.00×10^2 Ω

28. (a) 1.50×10^2 V

(b) 1.50×10^5 mV

30. (a) 9.80×10^2 kHz

(b) 9.80×10^{-1} MHz

PROBLEMS 7-2

2. (a) 3.8×10^{-3} km

(b) 12.5 ft

(c) 4.16 yd

4. (a) 6.48 km

(b) 4.02 mi

(c) 648×10^3 cm

6. 26.8 m/s

8. 100 mm

10. 4.98 pF/m

12. 2.1×10^{-3} dB/cm

14. 2.24×10^5 mi/h

PROBLEMS 7-3

2. 2.49 Ω **4.** 170 Ω **6.** 62 MHz

8. Solve for equation $\delta = \dfrac{6.63 \times 10^{-3}}{\sqrt{f}}$ cm for depth in cm

When f is in MHz:

$$\delta = \frac{6.63 \times 10^{-3}}{\sqrt{0.6}} \text{ cm} \times \frac{10 \text{ mm}}{1 \text{ cm}}$$

$$\underline{\delta = 8.56 \times 10^{-2} \text{ mm}}$$

10. $R_{ac} = 9.98 \times 10^{-4} \dfrac{\sqrt{f}}{d}$ Ω/ft f in MHz, d in inches

$$R_{ac} = 9.98 \times 10^{-4} \frac{\sqrt{f}}{d \text{ cm}} \times \frac{1 \text{ ft}}{12 \text{ in}} \times \frac{1 \text{ in}}{2.54 \text{ cm}} \text{ Ω/cm}$$

$$R_{ac} = 83.2 \times 10^{-6} \frac{\sqrt{f}}{d} \text{ Ω/cm}$$

With f in MHz and d in cm

$$R_{ac} = 8.32 \times 10^{-5} \frac{\sqrt{120}}{0.079 \text{ mm} \times \dfrac{1 \text{ cm}}{10 \text{ mm}}} \text{ Ω/cm}$$

$$\underline{\therefore R_{ac} = 0.1153 \text{ Ω/cm}}$$

12. $\dfrac{\lambda}{c} = \dfrac{1}{f}, \ \dfrac{\lambda}{1} = \dfrac{c}{f}, \ m, \ \lambda = \dfrac{300 \times 10^{6} \times 10^{2}}{f \times 10^{3}}$ cm

$$\therefore \lambda = \frac{3 \times 10^{7}}{f} \text{ cm (when } f \text{ is in kHz)}$$

14. $\lambda = \dfrac{3 \times 10^{4}}{f}$ cm (when f is in MHz)

$$\frac{\lambda}{2} = \frac{1.5 \times 10^{4}}{f} \text{ cm} \qquad L_D = \frac{\lambda}{2} - 6\% \text{ cm}$$

$$L_D = \frac{0.94 \times 1.5 \times 10^{4}}{f} \text{ cm} \qquad \therefore L_D = \frac{1.41 \times 10^{4}}{f} \text{ cm}$$

16. $L_r = L_D + 5\%$ cm   5% of $1.41 \times 10^{4} = 0.0705 \times 10^{4}$

$$\therefore L_r = \frac{1.48 \times 10^{4}}{f} \text{ cm}$$

18. $L_d = L_D - 5\%$ cm $\therefore L_d = \dfrac{1.34 \times 10^{4}}{f}$ cm

20. $680 \times 10^{6} \times 0.25 \times 10^{-6} = V = 1.70 \times 10^{2}$ V

PROBLEMS 8-1

2. 530 Ω **4.** 80 V **6.** 44.0 V **8.** 24 V

10. (a) 9.3 mA (b) 75.3 Ω (c) 41.7 Ω

(d) Diode resistance varies with changes in current.

PROBLEMS 8-2

2. (a) 32.6×10^3 W

(b) $\dfrac{32.6 \times 10^3}{746} = 43.70$ HP

4. 1.72 kW **6.** 432 mW **8.** 1 kW **10.** 0.161 W

12. (a) 178 mW (b) 19.8 V (c) -19.8 V **14.** (a) 6 mA (b) 36 mW

16. (a) 9.325 kW (b) 21.2 A (c) $39.16 **18.** (a) 78% (b) 5.26 kW

Solution for Prob. 20

(a) Motor power $= \dfrac{7.5 \times 746}{0.85} = \underline{6580 \text{ W}}$

(b) Power $= IV \quad \therefore I = \dfrac{6580}{230} = \underline{28.6 \text{ A}}$

(c) Total efficiency = 68%

$\therefore$ generator delivers 68% of 6.58 kW

$= \underline{4.48 \text{ kW}}$

(d) 4.48 kW = 2 kV × I $\quad \therefore \underline{I = 2.24 \text{ A}}$

(e) Total efficiency = 100 × 0.85 × 0.8 = $\underline{68\%}$

20. (a) 6.58 kW (b) 28.6 A (c) 4.48 kW (d) 2.24 A (e) 68%

DIAGRAM FOR PROBLEMS 8-3. (Problems begin on p. 12)

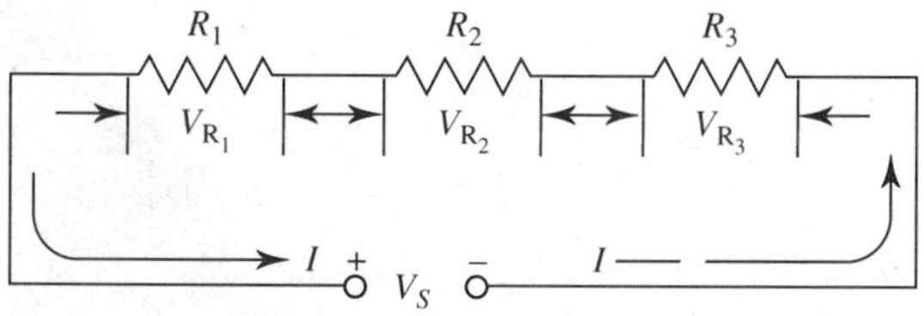

circuit diagram for problems 2 and 6

PROBLEMS 8-3. (See diagram on p. 11)

2. $R_1 = 1.8\ \text{k}\Omega$, $R_2 = 4.7\ \text{k}\Omega$, $R_3 = 1.2\ \text{k}\Omega$, $V_S = 100\ \text{V}$

$R_t = R_1 + R_2 + R_3 = 7.70\ \text{k}\Omega$

(a) $I = \dfrac{V_S}{R_t} = \dfrac{100}{7.7} \times 10^{-3} = \underline{12.98\ \text{mA}}$

(b) $V_{R_1} = IR_1 = \underline{23.4\ \text{V}}$ Check, does $V_S = V_{R_1} + V_{R_2} + V_{R_3}$?

$V_{R_2} = IR_2 = \underline{61.0\ \text{V}}$

$V_{R_3} = IR_3 = \underline{15.6\ \text{V}}$

4. (a) $I = \dfrac{P}{V} = \dfrac{240}{220} = \underline{1.09\ \text{A}}$

(b) Voltage across each lamp $\dfrac{220}{4} = \underline{55\ \text{V}}$

(c) $P_1 = 1.09^2 \times 50.4 = 59.9\ \text{W} \approx \underline{60\ \text{W}}$

6. (a) $I = \dfrac{470}{R_t}$, $V_{R_3} = 470 - 127 = 343\ \text{V}$

$I = \dfrac{V_{R_3}}{R_3} = \dfrac{343}{150 \times 10^3} = \underline{2.29\ \text{mA}}$

(b) $R_1 = \dfrac{76}{2.29} \times 10^3 = \underline{33.2\ \text{k}\Omega}$

(c) $R_2 = \dfrac{51}{2.29} \times 10^3 = \underline{22.3\ \text{k}\Omega}$

(d) $P_{R_1} = 2.29 \times 10^{-3} \times 76 = \underline{174\ \text{mW}}$

$P_{R_2} = 2.29 \times 10^{-3} \times 51 = \underline{117\ \text{mW}}$

$P_{R_3} = 2.29 \times 10^{-3} \times 343 = \underline{785\ \text{mW}}$

8.

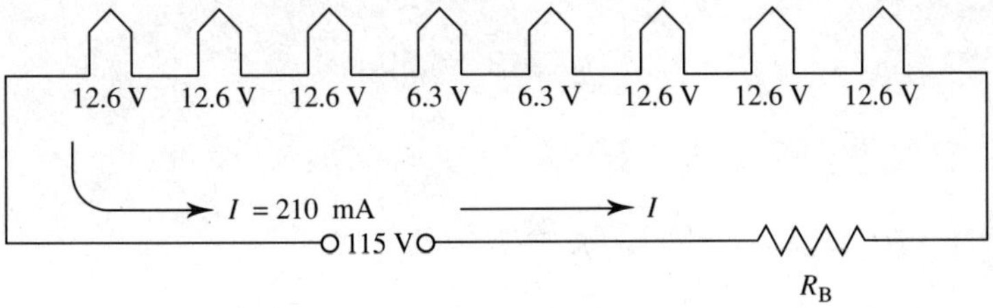

Total filament voltage = 6 × 12.6 + 2 × 6.3 = 88.2 V

$V_{R_B} = 115 - 88.2 = 26.8$ V, $I_B = I = 210$ mA

$$\therefore R_B = \frac{V_{R_B}}{I_B} = \frac{26.8}{210 \times 10^{-3}} = \underline{127.6\ \Omega}$$

10.

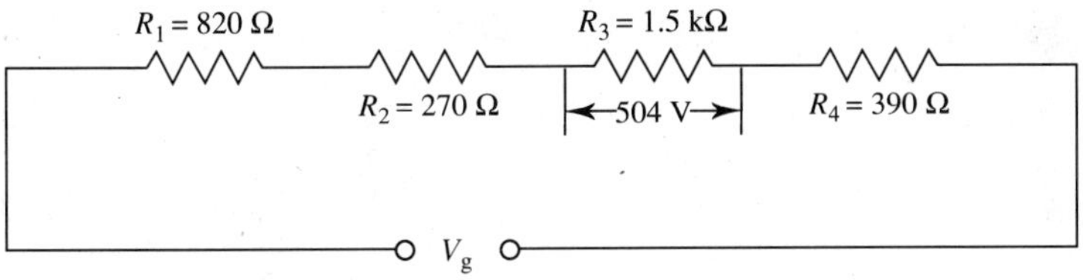

(a) $I = \dfrac{504}{1.5} \times 10^{-3} = 336$ mA

$V_g = I(R_1 + R_2 + R_3 + R_4) = 336 \times 10^{-3}(2980)$

$\therefore V_g = \underline{1\ \text{kV}}$

(b) $P = I^2R$ $\quad \therefore P_{R_1} = \underline{92.6\ \text{W}}$ $\quad P_{R_2} = \underline{30.5\ \text{W}}$

$P_{R_3} = \underline{169\ \text{W}}$ $\quad P_{R_4} = \underline{44\ \text{W}}$

Check that $IV = P_{R_1} + P_{R_2} + P_{R_3} + P_{R_4} = P_t$ W

PROBLEMS 8-4

2.

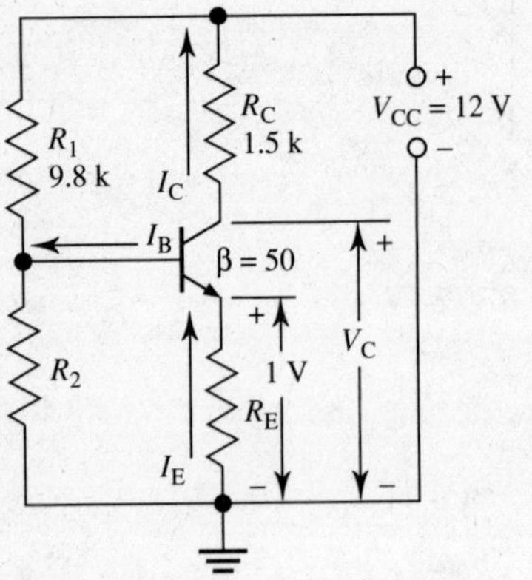

Data: $I_B = 0.1$ mA, $\beta = 50$

$V_{BE} = 0.2$ V

Solution: (a) $I_C = \beta I_B$

$= 50 \times 0.1 \times 10^{-3}$

$\underline{I_C = 5 \text{ mA}}$

$I_E = I_C + I_B = \underline{5.1 \text{ mA}}$

(b) Since $V_{BE} = 0.2$ V,

$V_{R_2} = V_{RE} + V_{BE} = 1.2$ V

$I_{R_2} = I_{R_1} - I_B$

and $I_{R_1} = \dfrac{12 - 1.2}{9.8 \times 10^3} = \dfrac{V_{R_1}}{I_1} = 1.1$ mA

$\therefore I_{R_2} = 1.1 - 0.1$ mA

$= 1$ mA

then $R_2 = \dfrac{V_{R_2}}{I_{R_2}} = \dfrac{1.2}{1 \times 10^{-3}} = \underline{1.2 \text{ k}\Omega}$

$R_E = \dfrac{V_{RE}}{I_E} = \dfrac{1}{5.1} \times 10^3 = \underline{196\ \Omega}$

(d) $V_C = V_{CC} - V_{RC}$, where $V_{RC} = I_C R_C$

$\therefore V_C = 12 - (1.5 \times \cancel{10^3} \times 5 \times \cancel{10^{-3}}) = 12 - 7.5 = \underline{4.5 \text{ V}}$

(a) $I_C = 5$ mA, $I_E = 5.1$ mA (b) $R_2 = 1.2$ kΩ, $R_E = 196$ Ω

(c) $I_{R_2} = 1$ mA

(d) $V_C = +4.5$ V

4. This circuit, shown in Fig. 8-24, was "solved" in Example 16, Sec. 8-9. The component and voltage values provide the following solution given that $I_G = 0$ A.

(a) $I_{R_1} = \dfrac{V_{R_1}}{R_1}$ from Problem 3 $\quad I_{R_1} = \dfrac{25}{11.5}\ \mu\text{A} = \underline{2.17\ \mu\text{A}}$

$\therefore V_{R_1} = 2.17 \times \cancel{10^{-6}} \times 10 \times \cancel{10^6} = \underline{21.7 \text{ V}}$

(b) $V_{R_2} = 2.17 \times \cancel{10}^{-6} \times 1.5 \times \cancel{10}^{6} = \underline{3.255\ \text{V}}$

Check $V_{R_1} + V_{R_2} = V_{DD} = 21.7 + 3.255 = 24.955$ V

which is -0.2%

(c) $V_{R_S} = V_S \qquad \therefore V_{GS} = V_{R_2} - V_{R_S}$

$= 3.255 - 6$

$V_{GS} = \underline{-2.74\ \text{V}}$

(d) R_D will dissipate most power (Largest current).

It is left as an exercise to prove this statement.

6.

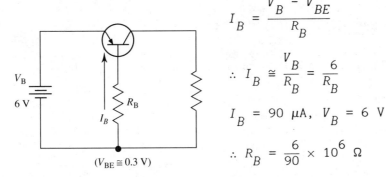

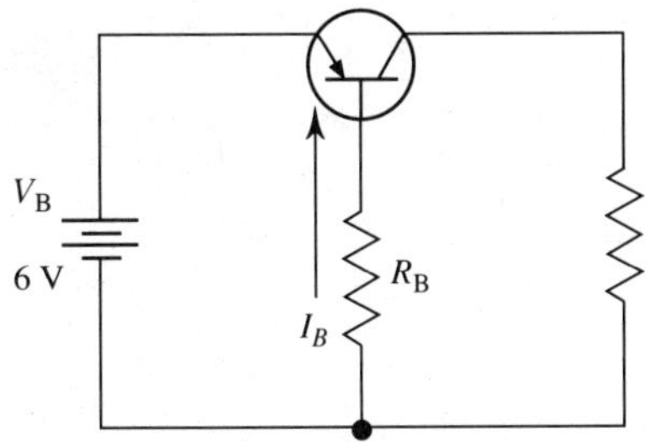

($V_{BE} \cong 0.3$ V)

$I_B = \dfrac{V_B - V_{BE}}{R_B}$

$\therefore I_B \cong \dfrac{V_B}{R_B} = \dfrac{6}{R_B}$

$I_B = 90\ \mu\text{A},\ V_B = 6$ V

$\therefore R_B = \dfrac{6}{90} \times 10^6\ \Omega$

$R_B = \underline{66.7\ \text{k}\Omega}$

8. $R_E = \dfrac{V_E}{I_E} = \dfrac{6 \times 10^3}{8} = \underline{750\ \Omega}$

10. (a) $R_E = \underline{1.2\ \text{k}\Omega}$

(b) $I_C = \alpha I_E + I_{CO} = 9.8 + 0.075 = \underline{9.875\ \text{mA}}$

(c) $I_E = I_C + I_B = 10$ mA $\quad \therefore I_B = \underline{125\ \mu\text{A}}$

PROBLEMS 9-1

2. (a) 84.2 Ω/km $\quad \therefore$ at 800 m $R = 84.2 \times 0.8 = \underline{67.4\ \Omega}$

(b) R at 1 m $= \dfrac{84.2}{1000} = \underline{0.0842\ \Omega}$

4. 1 km = 5.21 Ω $\quad$ (a) $\underline{0.005\ 21\ \Omega/\text{m}}$

(b) $0.005\ 21 \times 720 = \underline{3.75\ \Omega}$

(c) $0.005\ 21 \times 2200 = \underline{11.5\ \Omega}$

6. $\dfrac{R_1}{R_2} = \dfrac{A_2}{A_1} \qquad \therefore R_2 = \dfrac{R_1 A_1}{A_2} = \dfrac{0.0756 \times 0.5^2}{1.5^2} = \underline{0.0084\ \Omega}$

8. $\dfrac{32.1}{30} = 1.07\ \Omega/\text{m}, \qquad \therefore$ 702 Ω at a length of $\underline{656\ \text{m}}$

10. $\frac{1360}{1000} = 1.36\ \Omega/\text{m}$, $\frac{0.28}{1.36} = \underline{0.206\ \text{m}}$ or $\underline{206\ \text{mm}}$

PROBLEMS 9-2 [Use Table 9-1 for specific resistance values]

2. $R = \rho\frac{L}{A} = \frac{0.028\ 24 \times 120}{\pi(0.3215)^2} = \underline{10.4\ \Omega}$

4. $\underline{10.2\ \Omega}$

6. $\underline{40.6\ \Omega}$

8. $L = \frac{RA}{\rho} = \frac{100\pi\left(\frac{0.643}{2}\right)^2}{0.01771} = 1.834\ \text{km}$

$\therefore \Omega/\text{km} = \frac{100\ \Omega}{1.834\ \text{km}} = \underline{54.5\ \Omega/\text{km}}$

10. $\ell = 2.38\pi(1.025)^2 = 444\ \text{m}$ $\therefore \Omega/\text{m} = \frac{2.38}{444}$

$\Omega/\text{m} = 0.00537 = 5.37\ \Omega/\text{km}$

$\therefore R$ for 2 km $= \underline{10.7\ \Omega}$

PROBLEMS 9-3

2. $[R_t = R_o(1 + \alpha t)]$

$\therefore R_{20} = 4.13(1 + 0.004\ 27 \times 20) = \underline{4.48\ \Omega}$

4. $3.32 = 3.05(1 + 0.003\ 93\ \Delta t)$

$\frac{3.32}{3.05} = 1 + 0.003\ 93\ \Delta t$ $\therefore \Delta t = \left(\frac{1}{0.00\ 393}\right)\left(\frac{3.32}{3.05} - 1\right)$

and $t = \Delta t + t = 22.53 + 20 = \underline{42.5^\circ\text{C}}$

PROBLEMS 9-4

2. (a) 3.71 mΩ
   (b) 133.9 g

4. (a) 8.62 km
   (b) 114 Ω

6. (a) 11.5 Ω
   (b) 0.624 kg

8. (a) 223 V
   (b) 96.9%

10. (a) No. 1 wire
    (b) 232.5 V

PROBLEMS 10-1

2. $\alpha^3\pi^3$

4. $\frac{\pi D^2}{4}$

6. $\frac{16}{9}\pi^2 R^6$

8. $\frac{9v^2}{i^2}$

10. $-\frac{27i^3r^3}{v^3}$

12. $1331\ X_L^6$

14. $\frac{V_s^2 N_p^2}{V_p^3}$

16. $\frac{14400f^2}{N^2}$

18. $-8\pi^3\ell^3L^3$

20. $\frac{125u^6v^9w^3x^{12}y^{15}}{512}$

PROBLEMS 10-2

2. $\pm a^2$

4. ± 25

6. $\pm 10mn^6$

8. $\pm 27\alpha^3$

10. $-5x$

12. $\pm 20\pi\ell L$

14. $2\lambda\psi^2$

16. $\pm 11x^5y^6z^3$

18. $\pm\frac{5m^2np^4}{8a^2bc^3}$

20. $-\frac{2\pi X_L}{3Z^2X_C^4}$

22. $\pm\frac{14hn^2p^3}{11ab^2c}$

24. $-\frac{x^2y^4z^5}{3}$

PROBLEMS 10-3

2. $\frac{1}{2}\left(m + \frac{1}{2}q\right)$

4. $\frac{xy}{36}(9x^2 - 12y + 4xy)$

6. $2V(6I - 5R)$

8. $4\omega X_L(\omega^3L^3 - 3 + 7\omega X_L)$

10. $\frac{IRZ}{48}(12IRZ - 4Z + 3I)$

12. $27\theta\lambda\phi(2\theta^2\lambda + 3\theta\phi^2 - 4\lambda^2\phi)$

14. $\frac{X_L^2X_C^2}{72}(2X_L - 4X_C + 1)$

PROBLEMS 10-4

2. $C^2 + 14C + 49$

4. $R^2 - 4R + 4$

6. $m^2 - 6m + 9$

8. $4f^2 + 20fg + 25g^2$

10. $4\alpha^2 - 12\alpha\beta + 9\beta^2$

12. $4\lambda^2 - 20\lambda\mu + 25\mu^2$

14. $m^4 + 12m^2 + 36$

16. $4\theta^2 - 52\theta^2\phi + 169\phi^2$

18. $900 + 120 + 4 = 1024$

20. $900 + 300 + 25 = 1225$

22. $4\pi^2f^2L_1^2 - 4\pi fL_1Z + Z^2$

24. $\frac{R_1^2}{4} + \frac{R_1R_2}{4} + \frac{R_2^2}{16}$

26. $\frac{\phi^6\lambda^2}{9} + \frac{\alpha^2\phi^3\lambda}{3} + \frac{\alpha^4}{4}$

28. $y^2 + 4y + 4$

30. $\alpha^2 + \frac{2\alpha}{3} + \frac{1}{9}$

32. $\mu^2 - \frac{\mu}{6} + \frac{1}{144}$

34. $X_1^4 + \frac{4X_1^2}{3} + \frac{4}{9}$

36. $\frac{X^2}{4} + XY + Y^2$

38. $4 + 28xy + 49x^2y^2$

40. $4\phi^2 + 3\phi\theta^2 + \frac{9\theta^4}{16}$

PROBLEMS 10-5

2. $6x$

4. $2CR$

6. $20X_C$

8. $80L_1M$

10. y^2

12. $49\theta^2$

14. $36X^2$

16. $\frac{\theta^2}{4}$

18. $\frac{X^2}{36}$

20. $\pm(p - 6q)$

22. $\pm(V + 6I)$

24. $\pm(8\omega\lambda + \Omega^2)$

26. $\pm\left(2L_1 - \frac{3}{5}L_2\right)$

28. $\pm\left(\frac{M^2}{3} - \frac{2Z^4}{9}\right)$

PROBLEMS 10-6

2. $5ac(2b + 7d)$

4. $5i^2(V^2 - 2IRV + I^2R^2) = 5i^2(V - IR)^2$

6. $\frac{2}{r}\left(\frac{V^4}{3} + 2V^2v^2 + 3v^4\right) = \frac{2}{3r}(V^4 + 6V^2v^2 + 9v^4) = \frac{2}{3r}(V^2 + 3v^2)^2$

8. $\frac{2}{5\phi\omega}\left(9R^2 + 3R\ell L + \frac{1}{4}\ell^2L^2\right) = \frac{2}{5\phi\omega}\left(3R + \frac{1}{2}\ell L\right)^2$

10. $\frac{8F}{C}(25\ell^2 + 60\ell x + 36x^2) = \frac{8F}{C}(5\ell + 6x)^2$

PROBLEMS 10-7

2. $\lambda^2 - 9$

4. $X_L^{\ 2} - X_C^{\ 2}$

6. $4\pi^2R_1^{\ 2} - 4\pi^2R_2^{\ 2} = 4\pi^2(R_1^{\ 2} - R_2^{\ 2})$

8. $\frac{9}{16}\omega^4 - \frac{4}{25}\lambda^2$

10. $\frac{\theta^4}{4\phi^2} - \frac{9\alpha^2}{4\beta^2}$

PROBLEMS 10-8

2. $(Z + R)(Z - R)$

4. $(3m + 5p)(3m - 5p)$

6. $\left(\frac{\alpha}{\beta} - \frac{2\gamma}{3}\right)\left(\frac{\alpha}{\beta} + \frac{2\gamma}{3}\right)$

8. $\left(\frac{1}{V_1} - \frac{1}{v}\right)\left(\frac{1}{V_1} + \frac{1}{v}\right)$

10. $\left(\frac{1}{X_C} - \frac{V}{Q}\right)\left(\frac{1}{X_C} + \frac{V}{Q}\right)$

12. $(X + 2R)^2 - Z^2 = (X + 2R + Z)(X + 2R - Z)$

14. $16I^2 - \left(V^2 - \frac{14V}{X} + \frac{49}{X^2}\right)$

$= \left[4I - \left(V - \frac{7}{X}\right)\right]\left[4I - \left(V - \frac{7}{X}\right)\right]$

Answer $= \left(4I - V + \frac{7}{X}\right)\left(4I + V - \frac{7}{X}\right)$

PROBLEMS 10-9

2. $\theta^2 + 4\theta + 9$

4. $\lambda^2 - 7\lambda + 12$

6. $4r^2 + 10r + 6$

8. $25a^2 + 5a - 12$

10. $\frac{P^2}{4} + 4P + 12$

12. $\lambda^2 + \frac{17\lambda}{3} - 2$

14. $4\ell^2 + \frac{47\ell}{2} - 3$

16. $\frac{1}{R^2} + \frac{2}{R} - 3$

18. $v^2t^2 + \frac{5vt}{12} - \frac{1}{24}$

20. $M^2 - \frac{2L_1M}{X} - \frac{15L_1^2}{X^2}$

PROBLEMS 10-10

2. $(\phi + 2)(\phi + 7)$

4. $(\alpha - 2)(\alpha - 7)$

6. $(\theta + 6)(\theta + 3)$

8. $(\omega - 12)(\omega - 2)$

10. $(\alpha t - 9)(\alpha t - 4)$

12. $(g - 25)(g + 3)$

14. $(IR + 3V)(IR + V)$

16. $(\theta^2 - 6\phi)(\theta^2 + 2\phi)$

18. $\left(q^2 - \frac{1}{2}\right)\left(q^2 + \frac{1}{4}\right)$

20. $(v^2i^2 + 3p)(v^2i^2 - p)$

PROBLEMS 10-11

2. $I^2Z^2 - 2IZ - 15$

4. $10\psi^2 + 22\psi + 12$

6. $6A^2 - 6A - 12$

8. $21\theta^2 + 58\theta + 21$

10. $\frac{6}{\theta^2} - \frac{48}{\theta} + 72$

12. $I^2 - 12I - 108$

14. $36M^2 - 153M + 36$

16. $20 - 9m - 20m^2$

18. $20\theta^2 + 42\theta + 18$

20. $12x^2 + 13x - 35$

22. $\beta^2 + 0.1\beta - 0.3$

24. $I^2R^2 + 0.1IR - 0.9$

26. $72\delta^2 - \frac{18\delta}{3\eta} - \frac{8\delta}{2\eta} + \frac{2}{6\eta^2} = 72\delta^2 - \frac{10\delta}{\eta} + \frac{1}{3\eta^2}$

28. $120\pi^2L^2 - \frac{20\pi L}{3\pi C} - \frac{24\pi L}{3\pi C} + \frac{4}{9\pi^2C^2}$

$= 120\pi^2L^2 - \frac{44\not{\pi}L}{3\not{\pi}C} + \frac{4}{9\pi^2C^2} = \underline{120\pi^2L^2 - \frac{44L}{3C} + \frac{4}{9\pi^2C^2}}$

30. $\left(4m + \frac{3r}{p}\right)\left(6m - \frac{2r}{3p}\right)$

$= 24m^2 + \frac{18rm}{p} - \frac{8rm}{3p} - \frac{6r^2}{3p^2} = \underline{24m^2 + \frac{46mr}{3p} - \frac{2r^2}{p^2}}$

PROBLEMS 10-12

2. $(4\alpha + 1)(3\alpha + 2)$ **4.** $(R - 4)(2R - 3)$ **6.** $(2\beta - 5)(\beta - 2)$

8. $(9L_1 + 2)(2L_1 + 3)$ **10.** $(2P - 3W)(5P - W)$ **12.** $(5\lambda - 3\phi)(4\lambda - 2\phi)$

14. $(5I + 2R)(2I + 7R)$ **16.** $(6yz + 5w)(7yz - 4w)$ **18.** $(2)(\mu - 3\pi)(\mu + 3\pi)$

20. $(8m^2 - 9p)(3m^2 - 2p)$ **22.** $\left(\alpha - \frac{1}{2}\right)\left(\alpha - \frac{1}{3}\right)$ **24.** $\left(5Z - \frac{1}{4}\right)\left(2Z - \frac{1}{5}\right)$

PROBLEMS 10-13

2. $-125\alpha^6\beta^9\gamma^3$ **4.** $\pm 10\alpha\beta^3\gamma^2$ **6.** $-5a^3b^2c^4$

8. $\pm\frac{25I^2Rp}{8VW^3}$ **10.** $125a\left(\sqrt[3]{a^3x^3z^6}\right) = 125a^2xz^2$

12. $\pi(D_1 + D_2 + D_3)$ **14.** $0.12X(16C_1 - C_2 + 3C_3)$

16. $\frac{5}{12}\pi(d_1 - 3d_2 + 4d_3)$ **18.** $4Q^2 - 12P^2Q + 9P^4$

20. $0.04\alpha^2\lambda^2 - 0.12\alpha^2\lambda\mu + 0.09\alpha^2\mu^2$ **22.** $0.09V^2 + 0.3VIR + 0.25I^2R^2$

24. $20xy$ **26.** $4b^2 \quad \therefore (3a - 2b)^2$

28. $4L^4 \quad \therefore \left(2L^2 + \frac{3}{8}M\right)^2 = 4L^4 + \frac{3}{2}LM + \frac{9}{64}M^2$

30. $\pm(\alpha + 4\beta)$ **32.** $\pm\left(Y - \frac{B}{3}\right)$ **34.** $\pm\left(\frac{2}{5}V - 3X\right)$

36. $2m(3n - 5p)$ **38.** $5\pi(2\pi^2dr_1{}^2 + 7\pi dr_1r_2 + 5r_2{}^2)$

40. $\frac{12}{VI}(P^2 - 12PW + 36W^2) = \underline{\frac{12}{VI}(P - 6W)^2}$

42. $4\alpha^2\lambda^2 - 25\beta^2$ **44.** $64\theta^2 - 49\phi^2$ **46.** $0.9\varepsilon^2 - 0.25\eta^2$

48. $(3 + g_o)(3 - g_o)$ **50.** $\left(\frac{2}{5}\alpha\beta^2 - \frac{3}{4}\lambda\right)\left(\frac{2}{5}\alpha\beta^2 + \frac{3}{4}\lambda\right)$

52. $(0.1V + 0.6IR)(0.1V - 0.6IR)$

54. $2V + IR$

56. $0.2\alpha + 0.3\beta$

58. $x + y - 8$

60. $15 - 13G + 2G^2$

62. $0.03Z^2 + 0.08RZ - 0.6R^2$

64. $\frac{2}{3}\psi^2 + \frac{29}{21}\psi - \frac{5}{7}$

66. $4\pi^2 f^2 L^2 - 4\pi f L X_C - 3X_C^2$

68. $0.24x^2 + 1.5xy + 1.5y^2$

70. $\frac{v^2}{3} - \frac{6vs}{t} + \frac{24s^2}{t^2}$

72. $2(5X - 2)(3X + 1)$

74. $(m + 0.1)(m - 0.5)$

76. $\left(A + \frac{1}{8}\right)\left(A - \frac{1}{5}\right)$

78. $(3\alpha\beta\gamma - 5\Omega)(\alpha\beta\gamma + 2\Omega)$

80. $(g - 0.2h)(g + 0.3h)$

82. $\left(a + \frac{1}{b}\right)\left(a + \frac{1}{b}\right)$

84. $5\omega(\theta^2 - \phi^2) = 5\omega(\theta + \phi)(\theta - \phi)$

86. $\frac{2\lambda}{9}(36f + 1)(36f - 1)$

88. $\frac{x}{Z}\left(\frac{x^2}{9} - \frac{xy}{6} + \frac{y^2}{16}\right) = \frac{x}{144Z}(16x^2 - 24xy + 9y^2)$

$= \underline{\frac{x}{144Z}(4x - 3y)^2}$

90. $\frac{2V}{1350I}\left(450R_1^2 - 949R_1R_2 + 450R_2^2\right)$

$= \underline{\frac{2V}{1350I}(18R_1 - 25R_2)(25R_1 - 18R_2)}$

PROBLEMS 11-1

2. 20

4. $5\alpha^2$

6. $39x^2y^2z^3$

8. $2\alpha\beta^2\gamma$

10. $x + y$

12. $3\pi + \phi$

14. $3IR - 4V$

PROBLEMS 11-2

2. 462

4. $a^2b^3c^3$

6. $30\alpha^4\beta^3\gamma^3$

8. $51I^2R^2$

10. $(X - 4)(X - 5)(X - 6)$

12. $5(2 + 7\lambda)(2 - 7\lambda)^2$

14. $(4)(X_L + X_C)(X_L + 2X_C)(X_L + 3X_C)$

PROBLEMS 11-3

2. 20 **4.** $6\theta\omega$ **6.** $\alpha^2 - 2\alpha$ **8.** $\theta^2 + 8\theta + 7$ **10.** $x^2 - 4y^2$

12. $\frac{56}{128}$ **14.** $\frac{L^2 + 4L + 4}{L^2 - 4}$

PROBLEMS 11-4

2. $\frac{1}{6}$ **4.** $\frac{5}{18}$ **6.** $\frac{\theta}{4\phi^2}$ **8.** $\frac{2\lambda^2}{5\theta\phi}$ **10.** $\frac{15p + q}{5pq}$

12. $\frac{4(m - n)}{(m + n)(m - n)} = \frac{4}{m + n}$

14. $\frac{\cancel{(\alpha + 5\beta)}(\alpha - 2\beta)}{(2\alpha + \beta)\cancel{(\alpha + 5\beta)}} = \frac{\alpha - 2\beta}{2\alpha + \beta}$

PROBLEMS 11-5

2. $\frac{L_1L_2}{m - M}$ **4.** $\frac{\sqrt{L_1L_2}}{-\omega L}$ **6.** $\frac{\theta + \phi}{\tau - \rho}$ **8.** $\frac{-\mu V_s}{R_P + R_L}$ **10.** $\frac{\theta + \phi}{2\lambda^2}$

12. $\frac{1}{I + i}$ **14.** 1

16. $\frac{-(4st - v)\cancel{(st + v)}}{(2st - v)\cancel{(st + v)}} = \frac{v - 4st}{2st - v}$

PROBLEMS 11-6

2. $\frac{3i}{8}$ **4.** $\frac{IR - V}{I}$ **6.** $\frac{2A - 5}{A^2}$ **8.** $\frac{G(2Z + 1)}{2Z}$

10. $\frac{5x^2 - 10x + 5x - 30}{x^2 - 2x} = \frac{5(x^2 - x - 6)}{x^2 - 2x} = \frac{5(x - 3)(x + 2)}{x(x - 2)}$

12. $\frac{4c^2 - 4c - 8}{c^2} = \frac{4(c - 2)(c + 1)}{c^2}$

14. $\frac{2a + 2b - a + b}{8} = \frac{a + 3b}{8}$

16. $\frac{3x(x - 1) - 2(x + 1)(x - 1)}{6x^2} = \frac{3x^2 - 3x - 2x^2 + 2}{6x^2}$

Answer $= \frac{(x - 1)(x - 2)}{6x^2}$

18. $\frac{\cancel{3P^2} - 75 - \cancel{3P^2} + 15P}{P^2 - 25} = \frac{15P - 75}{P^2 - 25} = \frac{15(P - 5)}{(P + 5)(P - 5)} = \underline{\frac{15}{P + 5}}$

20. $\frac{(5\omega - 3\pi)(3\omega + 5\pi) - 50\omega\pi + 30\pi^2}{(5\omega - 3\pi)(3\omega + 5\pi)} = \frac{15\omega^2 - 34\omega\pi + 15\pi^2}{(5\omega - 3\pi)(3\omega + 5\pi)} = \underline{\frac{3\omega - 5\pi}{3\omega + 5\pi}}$

22. $8\frac{2}{5}$

24. $8\alpha - 4 + \frac{1}{\alpha}$

26.

$$
\begin{array}{r|l}
 & G + 7 \\
\cline{2-2}
G - 1 & G^2 + 6G + 8 \\
 & \underline{G^2 - G} \\
 & \quad 7G + 8 \\
 & \quad \underline{7G - 7} \\
 & \qquad +15 \quad \text{remainder}
\end{array}
$$

$$\text{Answer} = G + 7 + \frac{15}{G - 1}$$

28.

$$
\begin{array}{r|l}
 & 3\phi^3 + \phi^2 - 2\phi - 5 \\
\cline{2-2}
2\phi^2 - \phi + 3 & 6\phi^5 - \phi^4 + 4\phi^3 - 5\phi^2 - \phi + 20 \\
 & \underline{6\phi^5 - 3\phi^4 + 9\phi^3} \\
 & \quad 2\phi^4 - 5\phi^3 - 5\phi^2 \\
 & \quad \underline{2\phi^4 - \phi^3 + 3\phi^2} \\
 & \qquad -4\phi^3 - 8\phi^2 - \phi \\
 & \qquad \underline{-4\phi^3 + 2\phi^2 - 6\phi} \\
 & \qquad\quad -10\phi^2 + 5\phi + 20 \\
 & \qquad\quad \underline{-10\phi^2 + 5\phi - 15} \\
 & \qquad\qquad\qquad\qquad 35
\end{array}
$$

$$\text{Answer} = 3\phi^3 + \phi^2 - 2\phi - 5 + \frac{35}{2\phi^2 - \phi + 3}$$

30. $3x^2 - 4xy + 2y^2 - \frac{2y^4}{x + y}$

PROBLEMS 11-7

2. $\frac{105}{175}, \frac{56}{175}, \frac{80}{175}$

4. $\frac{q}{pq}, \frac{p}{pq}$

6. $\frac{\omega}{ir\omega}, \frac{ir}{ir\omega}, \frac{i^2r}{ir\omega}$

8. $\frac{ML_2Q}{L_1L_2M}, \frac{L_1M}{L_1L_2M}, \frac{L_1L_2\sqrt{L_1L_2}}{L_1L_2M}$

10. $\frac{x^2 - xy}{xy - y^2}, \frac{2xy + y^2}{xy - y^2}$

12. $\frac{3\phi}{1 - \phi^2}, \frac{2 - 2\phi}{1 - \phi^2}, \frac{2 + 2\phi}{1 - \phi^2}$

14. $\frac{3(M - 1)}{6(M^2 - 1)}, \frac{10(M + 1)}{6(M^2 - 1)}, \frac{6(1 - 3M)}{6(M^2 - 1)}$

16. $\dfrac{R + 3Z}{(4R + 4Z)(R + 2Z)}, \dfrac{R + Z}{(2R + 4Z)(2R + 6Z)}, \dfrac{R + 2Z}{(R + 3Z)(R + Z)}$

$= \dfrac{R + 3Z}{4(R + Z)(R + 2Z)}, \dfrac{R + Z}{4(R + 3Z)(R + 2Z)}, \dfrac{R + 2Z}{(R + 3Z)(R + Z)}$

Lowest Common Denominator = $4(R + Z)(R + 2Z)(R + 3Z)$ [LCD]

$= \dfrac{(R + 3Z)^2}{4(R + Z)(R + 2Z)(R + 3Z)}, \dfrac{(R + Z)^2}{4(R + Z)(R + 2Z)(R + 3Z)},$

$\dfrac{4(R + 2Z)^2}{4(R + Z)(R + 2Z)(R + 3Z)}$

PROBLEMS 11-8

2. $\dfrac{7 - 20 + 6}{32} = \dfrac{-7}{32}$

4. $\dfrac{40X - 12X + 75X}{60} = \dfrac{103X}{60}$

6. $\dfrac{n - N}{Nn}$

8. $\dfrac{100\ell - 24\ell - 15\ell}{120} = \dfrac{61\ell}{120}$

10. $\dfrac{4L_2 + 3L_1 + 7}{L_1 L_2}$

12. $\dfrac{3\alpha^2 + 2\phi^2 + 6\lambda^2}{\alpha\phi\lambda}$

14. $\dfrac{3a + 12 - 7a + 7}{21} = \dfrac{19 - 4a}{21}$

16. $\dfrac{5 - 4}{2(\alpha + 3)} = \dfrac{1}{2(\alpha + 3)}$

18. $\dfrac{a(c - d) + b(c + d) + a(c + d) - b(c + d)}{c^2 - d^2} = \dfrac{2ac}{c^2 - d^2}$

20. $\dfrac{8\alpha - 16 - 2\alpha - 6}{(\alpha + 3)(\alpha - 3)(\alpha - 2)} = \dfrac{2(3\alpha - 11)}{(\alpha + 3)(\alpha - 3)(\alpha - 2)}$

22. $\dfrac{2\cancel{(R_1 + 1)}\cancel{(R_1 - 1)}(11R_1 - 2) - 3\cancel{(R_1 - 1)}\cancel{(R_1 + 1)}(5R_1 + 1)}{6(R_1 + 1)(R_1 - 1)\cancel{(R_1 - 1)}\cancel{(R_1 + 1)}}$

$= \dfrac{22R_1 - 4 - 15R_1 - 3}{6(R_1 + 1)(R_1 - 1)} = \dfrac{7\cancel{(R_1 - 1)}}{6(R_1 + 1)\cancel{(R_1 - 1)}} = \dfrac{7}{6(R_1 + 1)}$

24. $\dfrac{2(L - 2M)(L - M)^{\cancel{2}} - 2\cancel{(L - M)}(3M^2 - 3LM)}{2\cancel{(L - M)}(L - M)(L - M)}$

$= \dfrac{2L^2 - \cancel{2LM} - \cancel{4LM} + \cancel{4M^2} - \cancel{6}^{2}M^2 + \cancel{6LM}}{2(L - M)(L - M)}$

$= \dfrac{2L^2 - 2M^2}{2(L - M)(L - M)} = \dfrac{\cancel{2(L - M)}(L + M)}{\cancel{2(L - M)}(L - M)} = \dfrac{L + M}{L - M}$

26. $$\frac{2X_C(2X_C - 3X_L) - 3X_L(2X_C + 3X_L) + 8X_L^2}{4X_C^2 - 9X_L^2}$$

$$= \frac{4X_C^2 - 12X_LX_C - X_L^2}{4X_C^2 - 9X_L^2}$$

28. $$\frac{(a + b)(a - b) - (a - b)(a + b) + a - b}{a - b} = \frac{a + a^2 - b^2 - a^2 + b^2 - b}{a - b}$$

Answer = 1

30. $$\frac{\theta + 3\pi}{4(\theta + 2\pi)(\theta + \pi)} + \frac{\theta + 2\pi}{(\theta + 3\pi)(\theta + \pi)} - \frac{\theta + \pi}{4(\theta + 3\pi)(\theta + 2\pi)}$$

$$= \frac{(\theta + 3\pi)^2 + 4(\theta + 2\pi)^2 - (\theta + \pi)^2}{4(\theta + 2\pi)(\theta + \pi)(\theta + 3\pi)}$$

$$= \frac{\theta^2 + \overset{4}{\cancel{6}}\theta\pi + \overset{8}{\cancel{9}}\pi^2 + 4\theta^2 + 16\theta\pi + 16\pi^2 - \cancel{\theta^2} - \cancel{2\theta}\pi - \pi^2}{(4\theta^2 + 20\theta\pi + 24\pi^2)(\theta + \pi)}$$

$$= \frac{4\theta^2 + 20(\theta\pi) + 24\pi^2}{(4\theta^2 + 20(\theta\pi) + 24\pi^2)(\theta + \pi)} = \frac{1}{\theta + \pi} = \text{answer}$$

PROBLEMS 11-9

2. $\dfrac{\overset{1}{\cancel{2}} \times \overset{1}{\cancel{3}}}{\underset{1}{\cancel{2}} \times \underset{1}{\cancel{7}}} \times \dfrac{3}{\underset{1}{\cancel{3}} \times 5} \times \dfrac{\overset{1}{\cancel{5}} \times \overset{1}{\cancel{7}}}{2 \times \underset{1}{\cancel{5}}} = \dfrac{3}{10}$

4. $\dfrac{\overset{1}{\cancel{3}}}{\underset{1}{\cancel{5}}} \times \dfrac{\overset{2}{\cancel{10}}}{\underset{5}{\cancel{15}}} = \dfrac{2}{5}$

6. $-\dfrac{2}{3}\left(-\dfrac{5}{16} \times \dfrac{64}{15}\right) = -\dfrac{2}{3}\left(-\dfrac{4}{3}\right) = \dfrac{8}{9}$

8. $\dfrac{\overset{1}{\cancel{3}}r}{2\cancel{m}} \times \dfrac{\overset{5}{\cancel{15}}\cancel{m}r}{\underset{2}{\cancel{\cancel{18}}}} = \dfrac{5r^2}{4}$

10. $\dfrac{2L}{C}$

12. x

14. $\dfrac{(4\theta^2 - 1)(\theta - 4)}{(\theta^3 - 16\theta)(2\theta - 1)} = \dfrac{2\theta + 1}{\theta(\theta + 4)}$

16. $\dfrac{\cancel{(I - 2i)}\cancel{(I + 2i)}(2i)}{i\cancel{(I + 2i)}\cancel{(I - 2i)}} = 2$

18. $\dfrac{\cancel{(F + 1)}\cancel{(F + 1)}}{\cancel{P}\cancel{(P^2 - Z^2)}} \times \dfrac{\cancel{(P^2 - Z^2)}}{5F\cancel{(F + 1)}^2} \times \dfrac{\cancel{P}\cancel{(F - 5)}(F - 5)}{\cancel{(F - 5)}(F - 105)} = \dfrac{F - 5}{5F(F - 105)}$

20. $\dfrac{\cancel{(R - r)}\cancel{(R + r)}}{\cancel{r}\cancel{(r + R)}} \times \dfrac{R\cancel{(R - r)}}{(R - r)^{\cancel{2}}} \times \dfrac{\cancel{r}\cancel{(R - 2r)}}{\cancel{(R - r)}\cancel{(R - 2r)}} = \dfrac{R}{R - r}$

22. $\dfrac{(4I^2R-3)\cancel{(4I^2R+3)}}{\cancel{(4I^2R+3)}} \times \dfrac{\cancel{(I^2R-7)}\cancel{(I^2R+4)}}{\cancel{2}\cancel{(I^2R-4)}\cancel{(I^2R+4)}} \times \dfrac{\cancel{8}^{4}\cancel{(I^2R-4)}}{(8I^2R-6)\cancel{(I^2R-7)}}$

$= \dfrac{4(4I^2R-3)}{8I^2R-6} = \dfrac{16I^2R-(3\times 4)}{8I^2R-6} = \underline{2}$

24. $m - \dfrac{m^2}{m} = 0 \qquad \therefore \underline{\text{answer} = 0}$

26. $\dfrac{(I+3)\cancel{(I-2)}}{I^4-9I^2} \times \dfrac{(I-3)\cancel{(I-3)}}{\cancel{(I-2)}\cancel{(I+1)}} \times \dfrac{I^2\cancel{(I+1)}}{\cancel{(I-3)}} = \dfrac{I^4-9I^2}{I^4-9I^2} = \underline{1}$

28. $\dfrac{45+14\theta+\theta^2}{\cancel{\theta^2}} \times \dfrac{\cancel{3\theta^2}(\theta+2)}{(\theta+9)^{\cancel{2}}} \times \dfrac{\cancel{(\theta+4)}\cancel{(\theta+9)}}{\cancel{(\theta+4)}(\theta+5)} \times \dfrac{1}{\cancel{3}(\theta+1)}$

$= \dfrac{(\theta+9)(\theta+5)(\theta+2)}{(\theta+9)(\theta+5)(\theta+1)} = \dfrac{(\theta+2)}{(\theta+1)} \qquad \therefore \text{answer} = \dfrac{\theta+2}{\theta+1}$

30. $\left(\dfrac{1+2\ell+\ell^2}{\ell^2}\right)\left(\dfrac{\ell^3-\ell^2}{(\ell-6)(\ell+1)}\right)\left(\dfrac{2(\ell^2-1)-12\ell+2}{\ell^2-1}\right)$

$= \dfrac{\cancel{(\ell+1)}\cancel{(\ell+1)}}{\cancel{\ell^2}} \cdot \dfrac{\cancel{\ell^2}\cancel{(\ell-1)}}{(\ell-6)\cancel{(\ell+1)}} \cdot \dfrac{2\ell^2-\cancel{2}-12\ell+\cancel{2}}{\cancel{(\ell-1)}\cancel{(\ell+1)}} = \dfrac{2\ell^2-12\ell}{\ell-6} = \underline{2\ell}$

PROBLEMS 11-10

2. $\dfrac{33}{7}$

4. $\dfrac{\dfrac{XR-1}{X}}{\dfrac{XR+1}{X}} = \dfrac{XR-1}{XR+1}$

6. $\dfrac{i^2-64}{8} \times \dfrac{8}{8+i} = \dfrac{\cancel{(i+8)}(i-8)\cancel{(8)}}{\cancel{8}\cancel{(8+i)}} = \underline{i-8}$

8. $\dfrac{250\phi+2\lambda}{5\phi} \times \dfrac{5\theta}{250\phi+2\lambda} \qquad \therefore \underline{\text{answer}} = \dfrac{\theta}{\phi}$

10. $\dfrac{(\lambda+\pi)(\lambda+\pi)-\lambda^2-\pi^2}{\cancel{\lambda^3+\pi^2\lambda+\pi\lambda^2+\pi^3}} \times \dfrac{\cancel{\lambda^3+\pi^2\lambda+\pi\lambda^2+\pi^3}}{\lambda^2+\pi^2-\lambda(\lambda+\pi)}$

$= \dfrac{\cancel{\lambda^2}+2\pi\lambda+\cancel{\pi^2}-\cancel{\lambda^2}-\cancel{\pi^2}}{\cancel{\lambda^2}+\pi^2-\cancel{\lambda^2}-\lambda\pi} = \dfrac{2\lambda\cancel{\pi}}{\cancel{\pi}(\pi-\lambda)} \qquad \therefore \underline{\text{answer}} = \dfrac{2\lambda}{\pi-\lambda}$

12. $\dfrac{\omega^2+2\omega-15}{\omega} \div \dfrac{\omega^2-8\omega+15}{\omega^2} = \left(\dfrac{(\omega+5)\cancel{(\omega-3)}}{\cancel{\omega}}\right)\left(\dfrac{\omega^{\cancel{2}}}{(\omega-5)\cancel{(\omega-3)}}\right)$

$\underline{\text{Answer}} = \dfrac{\omega(\omega+5)}{\omega-5}$

14. $$\frac{\theta(\theta - \phi) - \theta(\theta + \phi)}{\theta^2 - \phi^2} \times \frac{\theta^2 - \phi^2}{\theta(\theta - \phi) + \theta(\theta + \phi)}$$

$$= \frac{\theta^2 - \theta\phi - \theta^2 - \theta\phi}{\theta^2 - \theta\phi + \theta^2 + \theta\phi} = -\frac{2\theta\phi}{2\theta^2} = -\frac{\phi}{\theta} = \underline{\text{answer}}$$

16. $$\frac{L_1}{Q - \cfrac{1}{\cfrac{Q^2 + 1}{Q}}} - \frac{L_1}{Q + \cfrac{1}{\cfrac{Q^2 - 1}{Q}}} = \frac{L_1}{Q - \cfrac{Q}{Q^2 + 1}} - \frac{L_1}{Q + \cfrac{Q}{Q^2 - 1}}$$

$$= \frac{L_1}{\cfrac{Q^3 + Q - Q}{Q^2 + 1}} - \frac{L_1}{\cfrac{Q^3 - Q + Q}{Q^2 - 1}} = \frac{L_1(Q^2 + 1) - L_1(Q^2 - 1)}{Q^3}$$

$$= \frac{\cancel{L_1Q^2} + L_1 - \cancel{L_1Q^2} + L_1}{Q^3} \qquad \therefore \underline{\text{answer} = \frac{2L_1}{Q^3}}$$

PROBLEMS 12-1

2. $\frac{2y}{8} - \frac{y}{8} = 3 \qquad \therefore \frac{y}{8} = 3$ and $y = 24$

4. $I - \frac{2I}{5} = \frac{1}{4} - \frac{1}{16} \qquad \therefore \frac{3I}{5} = \frac{3}{16}$ and $\underline{I = \frac{5}{16}}$

6. $\frac{1}{3} + \frac{Z}{5} - \frac{Z}{3} = 0 \qquad \therefore \frac{5 + 3Z - 5Z}{15} = 0,\ \frac{1}{3} = \frac{2}{15}Z \qquad \therefore \underline{Z = 2\frac{1}{2}}$

8. $\frac{6F + 2F - 6}{36} = \frac{9 + 9F}{36} \qquad \therefore \underline{F = -15}$

10. $\frac{24I + 18 - 3I + 15}{30} = \frac{10I}{30} \qquad \therefore \frac{33}{30} = \frac{-11I}{30} \qquad \therefore \underline{I = -3}$

12. $\frac{30x - 18 - 24x + 10x - 15 - 10x + 8}{30} = 0 \qquad \therefore \underline{x = 4\frac{1}{6}}$

14. $\frac{8(Z + 1) - 9(Z + 2) - 2(Z + 1)}{12} = 0 \qquad \therefore \frac{-3Z}{12} = \frac{12}{12} \qquad \therefore \underline{Z = -4}$

PROBLEMS 12-2

2. $V = 3$ 4. $\lambda = 5$ 6. $\phi = -2.4$

8. $18 - 4\alpha + 27 + 9\alpha - 3 + 16\alpha = 0 \qquad \therefore \underline{\alpha = -2}$

10. $\frac{2\theta - 10}{15} = \frac{0.6\theta + 3 - 1.8\theta + 1.2}{15} \qquad \therefore \underline{\theta = 4.44}$

12. $x = 120$ 14. $\omega = 3$

PROBLEMS 12-3

2. $x = 3$ **4.** $L = 6$ **6.** $R = 2$ **8.** $P = 2$ **10.** $W = 4$

12. $m = 3$ **14.** $R = 1$ **16.** $R = \frac{25}{4} = 6.25$

18. $$\frac{(2\phi + 7)(9\phi + 6) - [(3\phi - 5)(6\phi - 4)]}{54\phi^2 - 24} = \frac{6(17\phi + 7)}{54\phi^2 - 24}$$

$$18\phi^2 + 75\phi + 42 - 18\phi^2 + 42\phi - 20 = 102\phi + 42$$

$$117\phi - 102\phi = 20$$

$$15\phi = 20$$

$$\therefore \underline{\phi = \frac{4}{3}}$$

20. $$\frac{(a + 3)(a - 7)}{(a + 2)(a + 3)} - \frac{6(a + 2)}{(a + 2)(a + 3)} = \frac{(a - 7)(a + 6)}{(a + 2)(a + 3)}$$

$$(a + 3)(a - 7) = (a - 7)(a + 6) + 6(a + 2)$$

$$\cancel{a^2} - 4a - 21 = \cancel{a^2} - a - 42 + 6a + 12$$

$$30 - 21 = 6a + 4a - a$$

$$9 = 9a$$

$$\therefore \underline{a = 1}$$

22. Entire job = $\frac{6}{6}$ or $\frac{10}{10}$ and $nh = h$ when $n = 1$

$\therefore$ together: $\frac{n}{6} + \frac{n}{10} = 1$, $5n + 3n = 30$ $\therefore n = \frac{30}{8} = \underline{3.75 \text{ h}}$

24. See text.

26. Filling pipes minus emptying pipes = Full tank (time)

$\frac{n}{8} + \frac{n}{5} - \frac{n}{6} = 1$, $15n + 24n - 20n = 120$ $\therefore \underline{n = 6.32\ h}$

28. $x + y - z = 1/h$, $\frac{yz + xz - xy}{xyz} = 1/h$

$$\therefore \text{time} = \frac{xyz}{yz + xz - xy}\ h$$

30. and **32.** See text.

34. $\frac{900}{n + 2} = \frac{475}{n}$ $\therefore 425n = 950$; $\therefore \underline{n = 2.235\ h}$

36. $\frac{54 + d - 9}{d - 9} = 4$, $d = 27$ $\therefore$ fraction $= \frac{81}{27} = 3$

38. $\frac{n + 23}{n + 12} = \frac{4}{3}$ $\therefore 3n + 69 = 4n + 48$, $\underline{n = 21}$

40. $(\ell - 3)\left(\frac{2}{3}\ell + 3\right) = \left(\frac{2}{3}\ell\right)(\ell) + 3$

$$\frac{2}{3}\ell^2 + \ell - 9 = \frac{2}{3}\ell^2 + 3$$

$$\therefore \underline{\ell = 12 \text{ m}}, \text{ and } \underline{w = 8 \text{ m}}$$

PROBLEMS 12-4

2. $w_1 = w - aw + aw_2$, $w_2 = \dfrac{w(a - 1) + w_1}{a}$

4. $F = 1 + \dfrac{R_{eq}}{R}$, $R = \dfrac{R_{eq}}{F - 1}$

6. $I_1 = \dfrac{V_1 - I_2R + I_2s}{R}$ $\qquad \therefore I_1 = \dfrac{V_1 - I_2(R + s)}{R}$

$V_1 - I_1R - I_2R = I_2s$ $\qquad \therefore V_1 = I_1R + I_2(R + s)$

$\dfrac{I_2s - V_1}{I_1 + I_2} = -R$ $\qquad \therefore R = \dfrac{V_1 - I_2s}{I_1 + I_2}$

8. $I_{\lambda_2}R_b = V_{e_2} + V_\lambda - V_2$ $\qquad \therefore V_\lambda = I_{\lambda_2}R_b + V_2 - V_{e_2}$

and $V_2 = V_{e_2} + V_\lambda - I_{\lambda_2}R_b$

10. $\mu(g_m' - g_m) = g_m$

$\mu g_m' - \mu g_m = g_m$ $\qquad \therefore g_m' = \dfrac{g_m(1 + \mu)}{\mu}$ and $g_m = \dfrac{g_m'\mu}{1 + \mu}$

12. $\mu G_2 - \mu G_L = 2G_L + g_p - 2G_2$

$G_2(\mu + 2) = 2G_L + \mu G_L + g_p$ $\qquad \therefore G_2 = G_L + \dfrac{g_p}{\mu + 2}$

$G_2 = \dfrac{G_L(2 + \mu) + g_p}{\mu + 2}$ $\qquad \therefore G_L = G_2 - \dfrac{g_p}{\mu + 2}$

$G_L - G_2 = \dfrac{-g_p}{\mu + 2}$ $\qquad \therefore g_p = (G_2 - G_L)(\mu + 2)$

14. $\beta_m a + \beta_m b = m\pi a$

$\beta_m b = a(m\pi - \beta_m)$ $\qquad \therefore a = \dfrac{\beta_m b}{m\pi - \beta_m}$ and $b = \dfrac{a(m\pi - \beta_m)}{\beta_m}$

16. $I_n = \dfrac{\gamma I_p}{I - \gamma}$ $\qquad \therefore \gamma I_p = I_n(1 - \gamma)$ $\qquad \therefore I_p = \dfrac{I_n(1 - \gamma)}{\gamma}$ and $\gamma = \dfrac{I_n}{I_n + I_p}$

18. $Z_0(R\mu + R) + Z_0R_a = RR_a$

$Z_0R_\mu + Z_0R + Z_0R_a = RR_a$

$R_aR - Z_0R - Z_0\mu R = Z_0R_a$

$R(R_a - Z_0\mu - Z_0) = Z_0R_a$

$$\therefore R = \frac{Z_0R_a}{R_a - Z_0(\mu + 1)}$$

$$\therefore R_a = \frac{Z_0R(\mu + 1)}{R - Z_0}$$

$$\therefore \mu = \frac{R_a(R - Z_0)}{Z_0R} - 1$$

20. $B_C\sqrt{2}D + B_CF = \pi\sqrt{2}DF\ell_b$ and $B_C\sqrt{2}D = F(\pi\sqrt{2}D\ell_b - B_C)$

$D\sqrt{2}(\pi F\ell_b - B_C) = B_CF$ $\qquad$ $\underline{\therefore F} = \dfrac{B_C\sqrt{2}D}{\pi\sqrt{2}D\ell_b - B_C}$

$$\underline{\therefore D} = \frac{B_CF}{\sqrt{2}(\pi F\ell_b - B_C)}$$

22. $Z_1R_a + Z_1R = (\mu + 1)R_1R + R_aR_1 + RR_a$

$Z_1R_a - R_1R_a - RR_a = (\mu + 1)R_1R + Z_1R$

$R_a(Z_1 - R_1 - R) = (\mu + 1)R_1R - Z_1R$

$$\therefore R_a = \frac{(\mu + 1)R_1R - Z_1R}{Z_1 - R_1 - R}$$

Solve for R,

$R_a(Z_1 - R_1 - R) = R[R_1(\mu + 1) - Z_1]$

$R_a(Z_1 - R_1) = R[R_a + R_1(\mu + 1) - Z_1]$

$$\therefore R = \frac{R_a(Z_1 - R_1)}{R_a + R_1(\mu + 1) - Z_1}$$

24. $C_1R_1\left(\dfrac{R_2 - R_3}{R_3}\right) = \sqrt{2} - 2C_2R_3 \qquad \therefore R_1 = \dfrac{R_3\left(\sqrt{2} - 2C_2R_3\right)}{C_1(R_2 - R_3)}$

From solution for R_1, $C_1 = \dfrac{R_3\left(\sqrt{2} - 2C_2R_3\right)}{R_1(R_2 - R_3)}$

Solution for R_2:

$$2C_1R_3 = \sqrt{2} - \frac{C_1R_1R_2}{R_3} + \frac{C_1R_1\not{R_3}}{\not{R_3}}$$

$$\frac{C_1R_1R_2}{R_3} = \sqrt{2} + C_1R_1 - 2C_2R_3$$

$$\therefore R_2 = \frac{R_3\left(\sqrt{2} + C_1R_1 - 2C_2R_3\right)}{C_1R_1}$$

26. $r = \dfrac{\mu V_g - PR_P}{P}$, $rP + PR_P = \mu V_g \qquad \therefore P(r + R_P) = \mu V_g$

$$\therefore P = \frac{\mu V_g}{r + R_P}$$

Solve for μ from solution for P:

$$\therefore \mu = \frac{P(r + R_P)}{V_g}$$

28. $\ell_{out}(C_1 + C_2) = C_1\ell_{in} \qquad \therefore \ell_{in} = \ell_{out}\left(1 + \dfrac{C_2}{C_1}\right)$

Solve for C_1:

$$\ell_{out}C_2 = C_1\ell_{in} - C_1\ell_{out}$$

$$\ell_{out}C_2 = C_1(\ell_{in} - \ell_{out}) \qquad \therefore C_1 = \frac{C_2\ell_{out}}{\ell_{in} - \ell_{out}}$$

Solve for C_2 from C_1 solution:

$$\therefore C_2 = \frac{C_1(\ell_{in} - \ell_{out})}{\ell_{out}}$$

30. $\dfrac{GR_{pg}}{g_m R_{pg} - G} = r_p$, $r_p g_m R_{pg} - r_p G = GR_{pg}$

$$r_p g_m R_{pg} = G(r_p + R_{pg})$$

$$\therefore \underline{G} = \frac{r_p g_m R_{pg}}{r_p + R_{pg}} \text{ and } \underline{g_m} = \frac{G(r_p + R_{pg})}{R_{pg} + r_p}$$

32. $X[(\ell_1 - \ell_2) - (\ell_0 - \ell_2)] = K$

$$\cancel{-X\ell_2} + X\ell_1 - X\ell_0 + \cancel{X\ell_2} = K$$

$$X\ell_1 - X\ell_0 = K \qquad \therefore \underline{\ell_1} = \frac{K}{X} + \ell_0 \text{ and } \underline{K} = X(\ell_1 - \ell_0)$$

34. $\mu\beta = \dfrac{2N}{2L + N}$; $\mu\beta 2L + \mu\beta N = 2N$

$$2\mu\beta L = N(2 - \mu\beta) \qquad \therefore \underline{L} = \frac{N(2 - \mu\beta)}{2\mu\beta}$$

Transpose and solve for N: $\qquad \therefore \underline{N} = \dfrac{2\mu\beta L}{2 - \mu\beta}$

36. $n_2 = \dfrac{-n_1 h\nu}{kT} + n_1$, $n_2 = n_1\left(1 - \dfrac{h\nu}{kT}\right)$ or $\underline{n_2} = n_1\left(\dfrac{kT - h\nu}{kT}\right)$

$$\therefore \underline{n_1} = \frac{n_2 kT}{kT - h\nu}$$

38. $F = 1 + \dfrac{2T_s}{T_a X}$, $F - 1 = \dfrac{2T_s}{T_a X}$ $\qquad \therefore T_a = \dfrac{2T_s}{X(F - 1)}$

Solve for X and T_s: $X = \dfrac{2T_s}{T_a(F - 1)}$, $T_s = \dfrac{XT_a(F - 1)}{2}$

40. $\dfrac{C_3}{C_1 + C_2} = \dfrac{R_3}{\dfrac{R_1 + R_2}{R_1 R_2}}$ $\qquad \therefore C_3(R_1 + R_2) = R_1R_2R_3(C_1 + C_2)$

$$C_3(R_1 + R_2) = R_1R_2R_3C_1 + R_1R_2R_3C_2$$

$$\therefore \underline{C_2} = \frac{C_3(R_1 + R_2) - R_1R_2R_3C_1}{R_1R_2R_3}$$

Solve for R_1: $C_3R_1 + C_3R_2 = R_1R_2R_3(C_1 + C_2)$

$$C_3R_2 = R_1R_2R_3(C_1 + C_2) - C_3R_1$$

$$= R_1[R_2R_3(C_1 + C_2) - C_3]$$

$$\therefore \underline{R_1} = \frac{C_3R_2}{R_2R_3(C_1 + C_2) - C_3}$$

42. $\mu_0(\alpha - 1) = 1 + 1.5\frac{d_2}{d_1}$, $d_1[\mu_0(\alpha - 1) - 1] = 1.5d_2$

$$\therefore \underline{d_1} = \frac{1.5d_2}{\mu_0(\alpha - 1) - 1}$$

44. $Z_{am}^2X_s^2 + Z_{am}^2Z_{ab}^2 = R(X_P - X_s)Z_{ab}^2$

$Z_{am}^2X_s^2 = R(X_P - X_s)Z_{ab}^2 - Z_{ab}^2Z_{am}^2$ $\qquad \therefore \underline{Z_{ab}^2} = \dfrac{Z_{am}^2X_s^2}{R(X_P - X_s)Z_{am}^2}$

Solve for X_P: $Z_{am}^2(Z_{ab}^2 + X_s^2) = RZ_{ab}^2X_P - RZ_{ab}^2X_s$

$$\frac{Z_{am}^2Z_{ab}^2 + Z_{am}^2X_s^2 + RZ_{ab}^2X_s}{RZ_{ab}^2} = X_P$$

$$\therefore \underline{X_P} = \frac{Z_{ab}^2(Z_{am}^2 + RX_s) + Z_{am}^2X_s^2}{RZ_{ab}^2}$$

46. $\underline{R_b} = \dfrac{(C_g - C_{gf} - C_{gp})(r_p + r_b)}{\mu C_{gp}}$ and $\underline{C_{pg}} = \dfrac{(r_p + r_b)(C_g - C_{gf})}{r_p + r_b + \mu R_b}$

48. $K_\varepsilon^2 + \dfrac{K_\varepsilon^2(\tan^2 K_a)}{\varepsilon_p^2} = -a^2$

$K_\varepsilon^2 + a^2 = \dfrac{-K_\varepsilon^2(\tan^2 K_a)}{\varepsilon_p^2}$ $\qquad \therefore \underline{\varepsilon_p^2} = \dfrac{-K_\varepsilon^2(\tan^2 K_a)}{K_\varepsilon^2 + a^2}$

50. $VR_0 = I_2R_1R_0 + I_2R_1R_2 + I_2R_0R_2$

$VR_0 - I_2R_0R_2 = R_1(I_2R_2 + I_2R_0)$

$$\therefore R_1 = \frac{VR_0 - I_2R_0R_2}{I_2(R_0 + R_2)}$$

or $\underline{R_1} = \dfrac{R_0(V - I_2R_2)}{I_2(R_0 + R_2)}$

52. $V_b - V_c = \mu V_c + \mu V_s\left(\dfrac{R_p}{R_1 + R_p}\right)$

$$V_b - V_c - \mu V_c = \frac{\mu V_s R_p}{R_1 + R_p}$$

$$(R_1 + R_p)(V_b - V_c - \mu V_c) = \mu V_s R_p \qquad \therefore R_1 + R_p = \frac{\mu R_p V_s}{V_b - V_c - \mu V_c}$$

$$\therefore \underline{R_1} = \frac{\mu V_s R_p}{V_b - V_c(1 + \mu)} - R_p$$

Solve for V_s from R_1 solution:

$$R_1 + R_p = \frac{\mu V_s R_p}{V_b - V_c(1 + \mu)}$$

$$(R_1 + R_p)(V_b - V_c - \mu V_c) = \mu V_s R_p$$

$$\therefore \underline{V_s} = \frac{(R_1 + R_p)(V_b - V_c - \mu V_c)}{\mu R_p}$$

54. $\dfrac{2\ell s^2}{\alpha F N^2} = \dfrac{1}{\dfrac{F_2 + F_s}{F_2}} \quad \therefore \dfrac{s^2}{N^2} = \dfrac{\alpha F F_2}{2\ell F_2 + 2\ell F_s}$

$$s^2 2\ell F_2 + s^2 2\ell F_s = N^2 \alpha \ell F_2 \qquad \therefore s^2 2\ell F_s = N^2 \alpha F F_s - s^2 2\ell F_2$$

Solve for F_s and F_2: $\quad F_s = \dfrac{F_2 N^2 \alpha F}{2\ell s^2} - F_2$

$$F_2 = \frac{2\ell F_s s^2}{N^2 \alpha F - 2\ell s^2}$$

56. $T_m = \dfrac{T(\omega_{21}hv_{12})}{\omega_{32}kv_{12}T_m - \omega_{21}hv_{12}}$

$T_m^2\omega_{32}kv_{12} - T_m\omega_{21}hv_{12} = T(\omega_{21}hv_{12})$ $\quad\therefore T = \dfrac{T_m^2\omega_{32}k}{\omega_{21}h} - T_m$

$$\underline{T} = \frac{T_m^2\omega_{32}k - \omega_{21}hT_m}{\omega_{21}h}$$

Solve for h:

$$T\omega_{21}h + \omega_{21}hT_m = T_m^2\omega_{32}k$$

$$h\omega_{21}(T + T_m) = T_m^2\omega_{32}k \qquad \therefore \underline{h} = \frac{T_m^2\omega_{32}k}{\omega_{21}(T + T_m)}$$

58. $a_2 = \dfrac{FC}{\Omega_1\Omega_2 - \Omega_1 B - \Omega_2 B + B^2 + c^2}$ $\quad\therefore FC = a_2(\Omega_1\Omega_2 - \Omega_1 B - \Omega_2 B + B^2 + c^2)$

$\dfrac{FC}{a_2} + \Omega_2 B - B^2 - c^2 = \Omega_1\Omega_2 - \Omega_1 B,$ $\quad \Omega_1(\Omega_2 - B) = \dfrac{FC}{a_2} + \Omega_2 B - B^2 - c^2$

$\therefore \Omega_1 = \dfrac{FC}{a_2(\Omega_2 - B)} + \dfrac{a_2(\Omega_2 B - B^2 - c^2)}{a_2(\Omega_2 - B)},$ $\quad \Omega_1 = \dfrac{FC}{a_2(\Omega_2 - B)} + \dfrac{B(\Omega_2 - B) - c^2}{\Omega_2 - B}$

$$\therefore \Omega_1 = \frac{FC}{a_2(\Omega_2 - B)} - \frac{c^2}{(\Omega_2 - B)} + B$$

60. $i_s = \dfrac{v}{L\left(\dfrac{S_s L + R}{L}\right)} = \dfrac{v}{S_s L + R}$ $\quad\therefore S_s L + R = \dfrac{v}{i_s}$

$$\therefore \underline{L} = \frac{v - i_s R}{S_s i_s}$$

Solving for R: $\quad \underline{R} = \dfrac{v - i_s L S_s}{i_s}$

62. $\omega_{01}L = \dfrac{R_1R_2}{R_1 + R_2}$ $\quad\therefore L = \dfrac{R_1R_2}{\omega_{01}(R_1 + R_2)}$

Solving for R_1:

$$\omega_{01}LR_1 + \omega_{01}LR_2 = R_1R_2$$

$$\omega_{01}LR_2 = R_1R_2 - \omega_{01}LR_1$$

$$\omega_{01}LR_2 = R_1(R_2 - \omega_{01}L) \qquad \therefore \underline{R_1} = \frac{\omega_{01}LR_2}{R_2 - \omega_{01}L}$$

64. $\dfrac{HS^2R_cC + HS}{R_cC} = \dfrac{1}{C}$ $\quad \therefore HS^2R_cC + HS = R_c$

$$HS^2R_cC = R_c - HS$$

$$\therefore \underline{C} = \frac{R_c - HS}{HS^2R_c}$$

Solving for R_C from solution C:

$$HS^2R_CC = R_C - HS, \qquad \therefore HS = R_C(1 - HS^2C)$$

$$\therefore \underline{R_C} = \frac{HS}{1 - HS^2C}$$

66. $G_2p = \dfrac{A(p + \omega_1)(C_1 + C_2)}{(p + \omega)(C_1 + C_2) - ApC_2}$

$$G_2p = \frac{AC_1p + ApC_2 + A\omega_1C_1 + A\omega_1C_2}{pC_1 + pC_2 + \omega C_1 + \omega C_2 - A_pC_2}$$

$$G_2p^2C_1 + G_2p^2C_2 + G_2p\omega C_1 + G_2p\omega C_2 - G_2p^2AC_2 = ApC_1 + ApC_2 + A\omega_1C_1 + A\omega_1C_2$$

Gathering C_1 and C_2 terms and factoring:

$$C_2(G_2p^2 + G_2p\omega - G_2p^2A - Ap - A\omega_1) = C_1(Ap + A\omega_1 - G_2p^2 - G_2p\omega)$$

$$C_2 = \frac{C_1[A(p + \omega_1) - G_2p(p + \omega)]}{G_2p(p + \omega - pA) - A(p + \omega_1)}$$

68.

$$\frac{V_0}{V} = \frac{\dfrac{R_Bh_{fe} + R_B + h_{ie}}{R_B}}{\dfrac{R_BR_Eh_{fe} + R_ER_B + R_Eh_{ie} + R_Bh_{ie}}{R_BR_E}}$$

$$\frac{V_0}{V} = \frac{R_BR_Eh_{fe} + R_BR_E + R_Eh_{ie}}{R_BR_Eh_{fe} + R_BR_E + R_Eh_{ie} + R_Bh_{ie}}$$

$$V = \frac{V_0\left[R_BR_E\left(h_{fe} + 1\right) + h_{ie}\left(R_E + R_B\right)\right]}{R_E\left[R_B\left(h_{fe} + 1\right) + h_{ie}\right]}$$

Cross-multiply to remove fractions and solve for R_B.

Solution for R_B, from solution for V:

$$V_0R_BR_Eh_{fe} + V_0R_BR_E + V_0R_Eh_{ie} + V_0R_Bh_{ie} = VR_ER_Bh_{fe} + VR_BR_E + VR_Eh_{ie}$$

$$R_B(V_0R_Eh_{fe} + V_0R_E + V_0h_{ie} - VR_Eh_{fe} - VR_E) = R_Eh_{ie}(V - V_0)$$

$$R_B[V_0h_{ie} + R_E(V_0 - V)(h_{fe} + 1)] = h_{ie}R_E(V - V_0)$$

$$\therefore \underline{R_B} = \frac{h_{ie}R_E(V - V_0)}{V_0h_{ie} + R_E(V_0 - V)(h_{fe} + 1)}$$

70. $R_{in} = \dfrac{R_Eh_{fe}R_B + R_ER_B + R_Bh_{ie} + R_Eh_{ie}}{R_B + h_{ie}}$

$$R_{in}R_B + R_{in}h_{ie} = R_ER_Bh_{fe} + R_ER_B + R_Bh_{ie} + R_Eh_{ie}$$

$$R_B(R_{in} - R_Eh_{fe} - R_E - h_{ie}) = R_Eh_{ie} - R_{in}h_{ie}$$

$$R_B(R_{in} - h_{ie} - R_E(h_{fe} + 1)) = h_{ie}(R_E - R_{in})$$

$$\therefore \underline{R_B} = \frac{h_{ie}(R_E - R_{in})}{R_{in} - h_{ie} - R_E(h_{fe} + 1)}$$

Solve for R_E:

$$R_BR_{in} - R_Bh_{ie} + R_{in}h_{ie} = R_Eh_{ie} + R_BR_E(h_{fe} + 1)$$

$$R_BR_{in} - h_{ie}(R_B - R_{in}) = R_E[h_{ie} + R_B(h_{fe} + 1)]$$

$$\therefore \underline{R_E} = \frac{R_BR_{in} - h_{ie}(R_B - R_{in})}{h_{ie} + R_B(h_{fe} + 1)}$$

72.

$$MH = \frac{\dfrac{4\pi r^2}{T^2\pi - \cancel{2\alpha T^2} + \cancel{2\alpha T^2}}}{\pi - 2\alpha} \qquad \therefore MH = \frac{4\pi r^2(\pi - 2\alpha)}{T^2\pi}$$

$$MHT^2\pi = 4\pi^2r^2 - 8\pi r^2\alpha \qquad \alpha = \frac{4\pi^2r^2 - MHT^2\pi}{8\pi r^2}$$

$$\therefore \alpha = \frac{4\pi r^2 - MHT^2}{8r^2}$$

Solve for π from α solution:

$$(\alpha)8r^2 = 4\pi r^2 - MHT^2$$

$$4\pi r^2 = MHT^2 + \alpha 8r^2 \qquad \therefore \pi = 2\alpha + \frac{MHT^2}{4r^2}$$

74. $S_1 = \dfrac{Fd^2}{10S_2} = \dfrac{100(20)^2}{10 \times 50} = 80$ units

76. $Z_pZ_1 + Z_pZ_2 = Z_1Z_2$

$$Z_pZ_1 = Z_2(Z_1 - Z_p) \qquad \therefore \underline{Z_2} = \frac{Z_pZ_1}{Z_1 - Z_p}\ \Omega$$

78. $N_S = \dfrac{N_PV_S}{V_P} = \dfrac{400 \times 20}{100} \qquad \therefore \underline{N_S = 80 \text{ turns}}$

80. $(\text{F} - 32)\dfrac{5}{9} = \text{C} \qquad \therefore \text{C} = (77 - 32)\dfrac{5}{9} = \underline{25^\circ\text{C}}$

82. $L_t = L_0(1 + \alpha t), \qquad L_0 = \dfrac{L_t}{1 + \alpha t} = \dfrac{20}{1 + (24)(8.33 \times 10^{-2})}$

$$\therefore \underline{L_0 = 6.67}$$

84. $L = \dfrac{2P}{I^2}$

$$= \frac{2(1250)}{(2.5)^2} \qquad \therefore \underline{L = 400\ H}$$

86. $\dfrac{1}{R_P} = \dfrac{R_2R_3 + R_1R_3 + R_1R_2}{R_1R_2R_3} \qquad \therefore R_P = \dfrac{R_1R_2R_3}{R_2R_3 + R_1R_3 + R_1R_2}$

88. $\dfrac{1}{\ell} - \dfrac{1}{p} = \dfrac{1}{q} = \dfrac{p - \ell}{\ell p} \qquad \therefore q = \dfrac{\ell p}{p - \ell} = \dfrac{10 \times 40}{40 - 10}$

$$\therefore \underline{q = 13.3 \text{ cm}}$$

90. $I = \dfrac{V}{\dfrac{nR + r}{n}} = \dfrac{Vn}{nR + r}$

$$\therefore InR = Vn - Ir, \qquad Ir = Vn - InR$$

$$\therefore r = \frac{Vn - InR}{I}\ \Omega \text{ and } R = \frac{Vn - Ir}{In}\ \Omega$$

92. $Ir = n(V - IR) \qquad \therefore n = \dfrac{Ir}{V - IR} \qquad \therefore n = \dfrac{0.1I}{V - 32I}$ cells

94. From Prob. 92: $n = \dfrac{2 \times 4.5}{2.1 - (2 \times 0.6)} = \underline{10 \text{ cells}}$

96. $\text{Gain} = \dfrac{i_dr_d}{r_s + \dfrac{1}{g_m}} \qquad \therefore \text{Gain} = \dfrac{i_dr_dg_m}{r_sg_m + 1}$ and $i_d = \dfrac{\text{Gain}(r_sg_m + 1)}{r_dg_m}$

$$\therefore i_d = \frac{8.33[(400 \times 0.005) + 1]}{5 \times 10^3 \times 5 \times 10^{-3}} \cong 1\ A$$

*Note: Ideal FET $i_g = 0$ is considered for this problem; <u>not</u> practical unless input capacitance and Z_{in} is also calculated.

98. $IR_3 + I(R_1 + R_2) = V \qquad \therefore R_3 = \dfrac{V - I(R_1 + R_2)}{I}$

No! R_3 and I cannot be interchanged.

100. $g = \dfrac{2(S - v_0t)}{t^2} \qquad g = \dfrac{2(520.5 - 30)}{100} \qquad \therefore \underline{g = 9.81\ \text{m/s}^2}$

102. $\dfrac{a}{b} = \dfrac{a^2 - ab - x}{a^2 - ab + x} - 1, \qquad \dfrac{a}{b} = \dfrac{\not{a}^2 - \not{ab} - x - \not{a}^2 + \not{ba} - x}{a^2 - ab + x}$

$\therefore \dfrac{a}{b} = \dfrac{-2x}{a^2 - ab + x}, \qquad a^3 - a^2b + ax = -2bx$

$a^3 - a^2b = -x(2b + a)$

$\therefore x = \dfrac{a^2b - a^3}{2b + a} = \dfrac{(9)(4.62) - 27}{3 + (2)(4.62)} \qquad \therefore \underline{x = 1.19}$

104. From prob. 103: $\quad g_m = \dfrac{\Delta i_d}{v_{gs}}$

$v_{gs} = \dfrac{\Delta i_d}{g_m} = \dfrac{7.5 \times 10^{-\not{6}\,3}}{7.5 \times 10^{-\not{3}}} = 1 \times 10^{-3} \qquad \therefore v_{gs} = 1\ \text{mv}$

106. $\dfrac{V_t}{L} = I_1 - I_2$

$\therefore I_1 - I_2 = \dfrac{1800 \times 0.2}{6}$

$I_1 - I_2 = \underline{60\ \text{A}}$

108. $R_aR_1 + R_a(R_3 + R_2) = R_1R_3$

$R_1(R_a - R_3) = -R_a(R_3 + R_2)$

$\therefore R_1 = \dfrac{R_a(-R_3 - R_2)}{R_a - R_3}$

Substitute values: $R_1 = \dfrac{0.6(-2.14 - 3)}{0.6 - 2.14} = \dfrac{-3.08}{-1.54} = \underline{2\ \Omega}$

110. $\beta - \beta\alpha = \alpha \qquad \therefore \beta = \alpha + \beta\alpha, \ \beta = \alpha(1 + \beta)$

Solve for α: $\alpha = \dfrac{\beta}{\beta + 1} \qquad \therefore \alpha = \dfrac{284.7}{285.7} = \underline{0.996}$

PROBLEMS 13-1

2. $R = \dfrac{3600 \times 6800}{1040} \qquad \therefore \underline{R = 2.35\ \text{k}\Omega}$

4. 33 000 // 8 200

$= \underline{6.57\ \text{k}\Omega}$

6. $R_P = \frac{R^2}{2R} = \frac{1}{2}\ R\Omega$ (underlined)

8. From prob. 7: $V = 440$ V

$$P = \frac{V^2}{R} = \frac{120^2}{270} = \underline{53.3\text{ W}}$$

10. $$P_{R_1} = \frac{(220)^2}{12 \times 10^3} = \underline{4.03\text{ W}}$$

12. $I_t = I_{R_1} + I_{R_2}$ $\quad \therefore I_{R_1} = I_t - I_{R_2}$, $\quad \therefore I_{R_1} = (70.27 - 14.71) \times 10^{-3}$

if $I_{R_1} = 55.6$ mA and $V_g = I_{R_1} R_1 = 55.6 \times 10^{-3} \times 18 \times 10^3 = 1$ kV

$$R_2 = \frac{V_g}{I_{R_2}} = \frac{1000}{14.71 \times 10^{-3}} = \underline{68\text{ k}\Omega}$$

14. $P = IV = 70.27 \times 10^{-3} \times 1000 = \underline{70.27\text{ W}}$

PROBLEMS 13-2

2. $\frac{1}{R_P} = \frac{1}{180} + \frac{1}{470} + \frac{1}{680}$ $\quad \therefore R_P = \underline{109.2\ \Omega}$

4. $R_P = 24.3\ \Omega$

6. $R_P = 4.54\ \Omega$

8. (a) $R_P = \frac{R}{3} = \underline{33.3\text{ k}\Omega}$, (b) $R_P = \frac{R}{4} = 25\text{ k}\Omega$, (c) $= \underline{20\text{ k}\Omega}$

10. $I_t = I_1 + I_2 + I_3$, $I_1 = \frac{V}{R_1}$, $I_2 = \frac{V}{R_2}$, $I_3 = \frac{V}{R_3}$

$18.03 = \frac{475}{100} + \frac{475}{150} + I_3$, $\quad \therefore I_3 = 18.03 - 4.75 - 3.17$

$$\underline{I_3 = 10.11\text{ A}}$$

$$R_3 = \frac{V}{I_3} = \frac{475}{10.11} = \underline{47\ \Omega}$$

12. zero Ω

14. $R_2 \;//\; R_3 = 510 \;//\; 270 = 177\ \Omega$ *(0.18 kΩ to 2 significant figures)

$I_t = I_1 + I_2 + I_3$, $\quad I_2 + I_3 = I_t - I_1$

$= 4.38 - 1.52 = \underline{2.86\ \text{A}}$

$V = (I_2 + I_3)(R_2 \;//\; R_3)$

$= 2.86 \times 177 = \underline{506\ \text{V}}$

$I_1 = \dfrac{V}{R_1} \quad \therefore R_1 = \dfrac{V}{I_1} = \dfrac{506}{1.52} = \underline{333\ \Omega}$ *(0.33 kΩ)

*Answers for R_1 and $R_2 \;//\; R_3$ to two significant figures; $R_1 = 0.33$ kΩ

*These figures are not used in calculations. Introduced error would be more than 20%. Round off only for final answers.

16. $V_{R_1}\text{max} = \sqrt{100 \times 10 \times 10^3} = 1\ \text{kV}$. $\quad V_{R_2}\text{max} = \sqrt{50 \times 15 \times 10^3} = 866\ \text{V}$

$V_{R_3}\text{max} = \sqrt{10 \times 100 \times 10^3} = 1\ \text{kV}$

$\therefore$ (a) safe maximum voltage = 866 V

(b) $I_t = \dfrac{866}{R_1//R_2//R_3} = \dfrac{866}{5.66} \times 10^{-3} = I_t = \underline{153\ \text{mA}}$

PROBLEMS 13-3

2. $R_t = R_1 + \dfrac{R_3R_2}{R_2 + R_3} = 360 + \dfrac{(160)(470)}{470 + 160} = \underline{479.4\ \Omega}$

$I_t = 250.3\ \text{mA} \quad \therefore V_{R_3} = V_P = I_t \times R_P = 250.3 \times 10^{-3} \times 119.4 = \underline{29.9\ \text{V}}$

$P_{R_3} = \dfrac{\left(V_{R_3}\right)^2}{R_3} = \dfrac{(29.9)^2}{160} = \underline{5.58\ \text{W}}$

4. $I_t = \dfrac{V}{R_T} = \dfrac{120}{360 + 160} = \dfrac{120}{520} = \underline{231\ \text{mA}}$

6. From Prob. 5: $V_g = (R_1)(I_t) + V_{R_3} = 230\ \text{V}$

With R_3 short circuit: $I_t = \dfrac{230}{62} \times 10^{-3} = \underline{3.71\ \text{mA}}$

8. From Prob. 7: $R_2 = 6.8$ kΩ and $R_3 = 4.7$ kΩ

(a) $P_{R_2} = \dfrac{1000^2}{6800} = 147\ \text{W}$ $\qquad$ (b) $I_3 = \dfrac{1000}{4700} = \underline{213\ \text{mA}}$

10. $R_{ab} = R_1 + R_3//(R_2 + R_L) = 300 + 300//(900) = \underline{525\ \Omega}$

12.

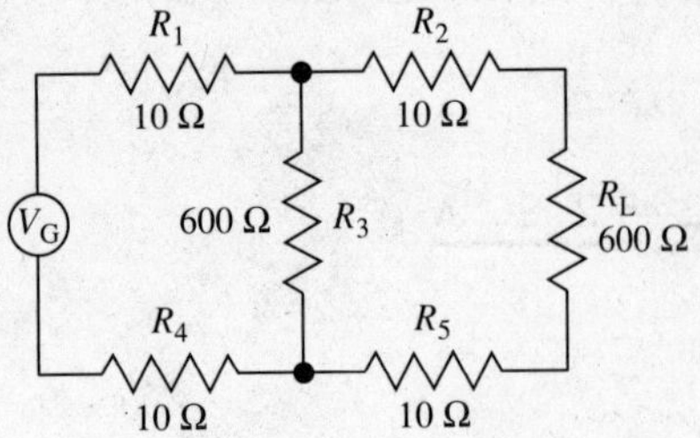

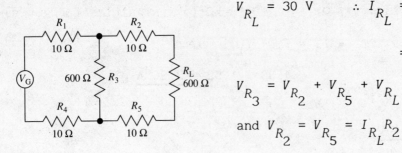

$V_{R_L} = 30 \text{ V} \quad \therefore I_{R_L} = I_{R_5} = I_{R_2}$

$= \frac{30}{600} = \underline{50 \text{ mA}}$

$V_{R_3} = V_{R_2} + V_{R_5} + V_{R_L}$

and $V_{R_2} = V_{R_5} = I_{R_L} R_2 = 0.5 \text{ V}$

$\therefore V_{R_3} = 31 \text{ V}, \ I_{R_3} = \frac{31}{600} = 51.7 \text{ mA}$

$I_t = I_{R_L} + I_{R_3} = (50 + 51.7)10^{-3} = \underline{102 \text{ mA}}$

14. $R_T = 10 + 14//5 + 3 + 22//(12 + 16//20)$

$= 13 + 3.684 + 22//20.889$

$= 16.684 + 10.715 \qquad \therefore R_T = 27.399 \ \Omega$ (27.4 Ω to 3 significant figures)

$I_T = \frac{V}{R_T} = \frac{100}{27.4} = \underline{3.65 \text{ A}}$

16. $R_T = 25//[(6//8//4) + 20 + (2//5//10)]$

$= 25//(1.846 + 20 + 1.25)$

$\therefore R_T = 25//23.1 = 12 \ \Omega \qquad \therefore I = \frac{230}{12} = \underline{19.2 \text{ A}}$

18. $R_T = 25//(1.846 + 1.25)$

$= 2.755 \ \Omega$ (Note: 2.76 Ω gives 19.1667 kW; 19.2 to three significant figures)

$P = \frac{V^2}{R} = \frac{230^2}{2.755} = \underline{19.2 \text{ kW}}$

PROBLEMS 14-1

2. $V_{fsd} = 100 \times 10^{-6} \times 1250 = 125 \text{ mV}, \ I_{\text{SHUNT}} = 9.9 \text{ mA}$

$\therefore R_{\text{SHUNT}} = \frac{125 \text{ mV}}{9.9 \text{ mA}} = \underline{12.6 \ \Omega}$

4. $V_{fsd} = 1 \times 10^{-3} \times 42 = 42 \text{ mV}, \qquad V_{\text{SHUNT}} = 42 \text{ mV}$

$I_{\text{SHUNT}} = \frac{V_{fsd}}{R_{\text{SHUNT}}} = \frac{42 \times 10^{-3}}{0.4719} = 89 \text{ mA}$

$I_{\text{meter}} + I_{\text{SHUNT}} = f.s.d. = 90 \text{ mA} \qquad \therefore$ <u>multiply by 90</u>

6. $R_1 + R_2 + R_3 + R_4 + R_5 = 1.1\ k\Omega$ and $N = \dfrac{fsd \text{ meter}}{fsd \text{ SHUNT}}$

$$R_{a-e} = \frac{R_{a-f} + R_m}{N} = \frac{2200}{20} = 110\ \Omega \qquad \therefore R_{a-f} = 110\ \Omega$$

Since $R_{a-e} = R_{a-f} = 110\ \Omega$, $\underline{R_1 = 990\ \Omega}$

$$R_{a-d} = \frac{2200}{200} = 11\ \Omega,\ R_{a-e} - R_{a-d} = 99\ \Omega,\ \underline{R_2 = 99\ \Omega}$$

$$R_{a-c} = \frac{2200}{2000} = 1.1\ \Omega,\ R_{a-d} - R_{a-c} = 9.9\ \Omega,\ \underline{R_3 = 9.9\ \Omega}$$

$$R_{a-b} = \frac{2200}{20\ 000} = 0.11\ \Omega,\ R_{a-c} - R_{a-b} = 0.99\ \Omega,\ \underline{R_4 = 0.99\ \Omega}$$

$$R_5 = R_m - (R_1 + R_2 + R_3 + R_4) = 1100 - 1099.89 = \underline{0.11\ \Omega}$$

PROBLEMS 14-2

2. $R_m + R_{mp} = \dfrac{V_{fsd}}{I_{fsd}} = \dfrac{100}{50} \times 10^6 = 2\ \text{M}\Omega$

$$R_T = 2\ \text{M}//100\ \text{k} + 100\ \text{k} = 195.24\ \text{k}\Omega.\ I_T = \frac{75 \times 10^{-3}}{195.24}\ \text{mA}$$

$\therefore I_T = 0.38$ mA Solve for V_{R_1}:

(a) $V_{R_1} = I_T(R_m + R_{mp}//R_1) = 0.384 \times \cancel{10^{-3}} \times 95.24 \times \cancel{10^3} = \underline{36.6\ \text{V}}$

(b) $V_{AB} = 75$ V (c) $I_{R_2} = I_{R_1} = I_T = \dfrac{75}{200} \times 10^{-3} = \underline{375\ \mu\text{A}}$

PROBLEMS 15-1

2. $I_T = \dfrac{350}{27k + 68k + 75k} = \dfrac{350}{170} \times 10^{-3} = \underline{2.06\ \text{mA}}$

@ one end of 75 kΩ, $V = I_T(75k + 68k)$

$$= 2.06 \times 143 = \underline{294\ \text{V}}$$

@ other end of 75 kΩ, $V = 294 - 2.06(75)$

$$= 294 - 154$$

$$= \underline{140\ \text{V}}$$

$\therefore$ range of control voltage is $\underline{140\ \text{V to } 294\ \text{V}}$

4. 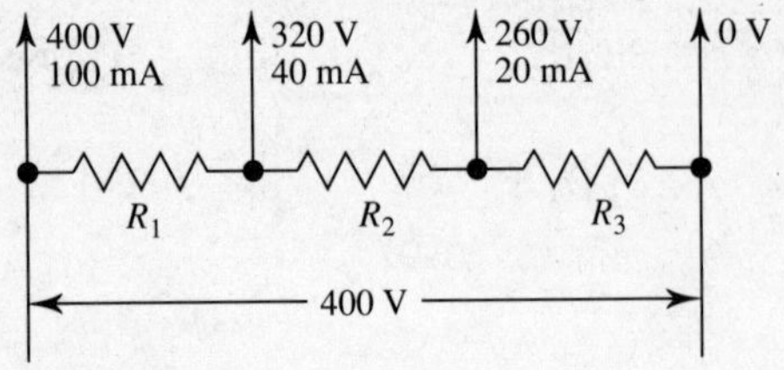

$I_L = 160$ mA, $I_B = 20$ mA

$\therefore I_T = 180$ mA (given)

$I_T = I_L + I_{bleeder}$

$I_{R_3} = 20$ mA, $V_{R_3} = 260$ V $\quad \therefore$ $\underline{R_3 = 13\ \text{k}\Omega}$

$I_{R_2} = 40$ mA, $V_{R_2} = 60$ V $\quad \therefore$ $\underline{R_2 = 1.5\ \text{k}\Omega}$

$I_{R_1} = 80$ mA, $V_{R_1} = 80$ V $\quad \therefore$ $\underline{R_1 = 1\ \text{k}\Omega}$

6. $I_L = 150$ mA, $\quad \therefore$ bleeder current $(I_B) = 20$ mA

$I_{R_1} = (63 + 42 + 20)$ mA $= 125$ mA, $V_{R_1} = 110$ V $\quad \therefore$ $\underline{R_1 = 880\ \Omega}$

(Use 910 Ω)

$I_{R_2} = 62$ mA, $V_{R_2} = 125$ V $\quad \therefore$ $\underline{R_2 = 2.02\ \text{k}\Omega}$ (2.2 kΩ preferred value)

$I_{R_3} = I_B = 20$ mA, $V_{R_3} = 225$ V $\quad \therefore$ $\underline{R_3 = 11.3\ \text{k}\Omega}$ (12 kΩ standard)

8. Total power $= V_g \times I_T$ W. $V_g = 460$ V and $I_T = I_L + I_B = 170$ mA

$\therefore$ total power = $\underline{78.2\ \text{W}}$

10. $I_{R_3} = I_B = 10$ mA, $V_{R_3} = 100$ V $\quad \therefore R_3$ Power = $\underline{1\ \text{W}}$ (Use 2 W)

$I_{R_2} = 30$ mA, $V_{R_2} = 90$ V $\quad \therefore R_2$ Power = $\underline{2.7\ \text{W}}$ (Use 5 W)

$I_{R_1} = 60$ mA, $V_{R_1} = 180$ V $\quad \therefore R_1$ Power = $\underline{10.8\ \text{W}}$ (Use 20 W)

$I_{R_4} = 100$ mA, $V_{R_4} = 50$ V $\quad \therefore R_4$ Power = $\underline{5\ \text{W}}$ (Use 10 W)

12. Supply voltage $= V_{R_1} + V_{R_2} + V_{R_3} + V_{R_4} = 420$ V

$\therefore$ voltage across A and B = $\underline{420\ \text{V}}$

PROBLEMS 15-2

2. (a) $I_1 = \dfrac{150}{R_1} = \dfrac{150}{2.2k} = \underline{68.2\ \text{mA}}$ $\quad$ (b) $I_2 = \dfrac{150}{4.7k} = \underline{31.9\ \text{mA}}$

4. $V_{SUPPLY} = I_1R_1 = I_2R_2 \qquad \therefore I_1R_1 = I_2R_2$

(a) $R_2 = \dfrac{I_1R_1}{I_2} = \dfrac{25 \times \cancel{10^{-3}} \times 200}{75 \times \cancel{10^{-3}}} = \underline{66.7\ \Omega}$

(b) $emf = 200 \times 25 \times 10^{-3} = \underline{5\ \text{V}}$

PROBLEMS 15-3

2. $R_x = \dfrac{R_1R_3}{R_2} = \dfrac{4.52}{100} = 0.0452\ \Omega$

4. $x = \dfrac{R_2L}{R_1 + R_2} = \dfrac{21.7 \times 2 \times 2.2}{33.3 + 21.7} = \underline{1.736\ \text{km}}$

6. $R_x = \dfrac{R_2R_L}{R_1 + R_2} = \dfrac{36.2 \times 5.62}{16.8 + 36.2} = \underline{3.838\ \Omega}$

No. 6 wire = 1.297 Ω/km $\qquad \therefore L = \dfrac{3.838}{1.297} = \underline{2.96\ \text{km}}$

PROBLEMS 16-1

2. The steepness of the *I-V* curve is inversely proportional to the resistance.

4. With constant applied voltage, current varies inversely as the value of the resistance.

6. (a) $\dfrac{96x}{60} + \dfrac{64x}{60} = 144 \qquad \therefore x = 54$ min

(b) Distance to meeting point, $\dfrac{54 \times 96}{60} = \underline{86.4\ \text{km}}$ and $\dfrac{54 \times 64}{60} = \underline{57.6\ \text{km}}$

(c) After 30 min, distance apart $= 144 - \left(\dfrac{96 + 64}{2}\right) = \underline{64\ \text{km}}$

8. $50 + 3x = y \qquad y = \$110$ and x = Number of radios

$5.5x = y \qquad \therefore x = 20$ when $y = \$110$

∴ Answer = after 21 radios are sold.

PROBLEMS 16-5

2. $y = 20x$

4. $s = u + qt$, $u = 0.4$, and $q = -0.000\ 805$

6. (a) $F = \frac{9}{5}C + 32$ (b) $25^\circ C = 77^\circ F$ (c) $C = (F - 32)\frac{5}{9} \qquad \therefore 165^\circ F = 74^\circ C$

8. (a) = 0.0924 *S* (b) = 0

PROBLEMS 17-1

2. $x = y = 5$

4. $x = 10,\ y = 8$

6. $m = 5,\ \ell = 40$

8. $F = \frac{1}{2},\ f = 4$

10. $P = 0.75,\ Q = 0.5$

PROBLEMS 17-2

2. $$\begin{array}{r} 2\alpha - 10\beta = 46 \\ \underline{2\alpha - 2\beta = 14} \\ -8\beta = 32 \end{array} \quad \therefore \beta = -4 \qquad \text{and} \qquad \begin{array}{r} 2\alpha - 10\beta = 46 \\ \underline{10\alpha - 10\beta = 70} \\ 8\alpha = 24 \end{array} \quad \therefore \alpha = 3$$

4. $$\begin{array}{r} 25V + 30I = 35 \\ \underline{12V + 30I = -30} \\ 13V = 65 \end{array} \quad \therefore V = 5 \qquad \text{and} \qquad \begin{array}{r} 10V + 12I = 14 \\ \underline{10V + 25I = -25} \\ 13I = -39 \end{array} \quad \therefore I = -3$$

6. $$\begin{array}{r} 5\theta + 4\phi = 12 \\ \underline{5\theta - 10\phi = 40} \\ 14\phi = -28 \end{array} \quad \therefore \phi = -2 \qquad \text{and} \qquad \begin{array}{r} 5\theta + 4\phi = 12 \\ \underline{2\theta - 4\phi = 16} \\ 7\theta = 28 \end{array} \quad \therefore \theta = 4$$

8. $$\begin{array}{r} 8\alpha - 4\beta = 12 \\ \underline{3\alpha + 4\beta = 10} \\ 11\alpha = 22 \end{array} \quad \therefore \alpha = 2 \qquad \text{and} \qquad \begin{array}{r} 6\alpha + 8\beta = 20 \\ \underline{6\alpha - 3\beta = 9} \\ 11\beta = 11 \end{array} \quad \therefore \beta = 1$$

10. $$\begin{array}{r} 36p - 27q = 45 \\ \underline{36p - 32q = 0} \\ 5q = 45 \end{array} \quad \therefore q = 9 \qquad \text{and} \qquad \begin{array}{r} 32p - 24q = 40 \\ \underline{27p - 24q = 0} \\ 5p = 40 \end{array} \quad \therefore p = 8$$

12. $$\begin{array}{r} 3Z_1 + Z_2 = 14 \\ \underline{3Z_1 + 6Z_2 = 39} \\ -5Z_2 = -25 \end{array} \quad \therefore Z_2 = 5 \qquad \text{and} \qquad \begin{array}{r} 6Z_1 + 2Z_2 = 28 \\ \underline{Z_1 + 2Z_2 = 13} \\ 5Z_1 = 15 \end{array} \quad \therefore Z_1 = 3$$

14. $$\begin{array}{r} 4I + 12i = 100 \\ \underline{4I + 16i = 124} \\ -4i = -24 \end{array} \quad \therefore i = +6 \qquad \text{and} \qquad \begin{array}{r} 4I + 12i = 100 \\ \underline{3I + 12i = 93} \\ I = 7 \end{array} \quad \therefore I = 7$$

16. $$\begin{array}{r} 15\alpha + 9\beta = -3 \\ \underline{15\alpha + 35\beta = 75} \\ -26\beta = -78 \end{array} \quad \therefore \beta = 3 \qquad \text{and} \qquad \begin{array}{r} 35\alpha + 21\beta = -7 \\ \underline{9\alpha + 21\beta = 45} \\ 26\alpha = -52 \end{array} \quad \therefore \alpha = -2$$

18. $$\begin{array}{rl} 0.27X_L + 0.012X_C &= 2.82 \\ 0.002X_L - 0.012X_C &= -0.1 \\ \hline 0.272X_L &= 2.72 \end{array} \quad \text{and} \quad \begin{array}{rl} 0.045X_L + 0.002X_C &= 0.47 \\ 0.045X_L - 0.27X_C &= -2.25 \\ \hline 0.272X_C &= 2.72 \end{array}$$

$$\therefore X_L = X_C = 10$$

20. $$\begin{array}{rl} 0.28L - 0.63X &= -1.12 \\ -0.54L + 0.63X &= -0.18 \\ \hline -0.26L &= -1.3 \end{array} \quad \therefore L = 5 \quad \text{and} \quad \begin{array}{rl} 0.24L - 0.54X &= -0.96 \\ -0.24L + 0.28X &= -0.08 \\ \hline -0.26X &= -1.04 \end{array} \quad \therefore X = 4$$

PROBLEMS 17-3

2. $a = 13 - 2b$

$$3(13 - 2b) + 15 = 3b$$

$$39 - 6b + 15 = 3b$$

$$54 = 9b \qquad \therefore \underline{b = 6} \text{ and } \underline{a} = 13 - 12 = \underline{1}$$

4. $\lambda = 17 - \mu$

$$17 - \mu + 8\mu = 122$$

$$7\mu = 105 \qquad \therefore \underline{\mu = 15} \text{ and } \underline{\lambda} = 17 - 15 = \underline{2}$$

6. $I_1 = \dfrac{74 - 7I_2}{5}$ and $I_2 = \dfrac{7I_1}{5}$

$$5I_2 - 7\left(\frac{74 - 7I_2}{5}\right) = 0 \quad \text{and} \quad 5I_1 + 7\left(\frac{7I_1}{5}\right) = 74$$

$$\frac{74I_2}{5} = \frac{518}{5} \qquad \frac{74I_1}{5} = 74$$

$$\therefore \underline{I_2 = 7} \text{ and } \underline{I_1 = 5}$$

8. $X_L = \dfrac{8X_C - 20}{3}$ and $X_C = \dfrac{44 - 8X_L}{3}$

$$3X_L - \frac{8}{3}(44 - 8X_L) = -20 \quad \text{and} \quad 3X_C - \frac{8}{3}(8X_C - 20) = 44$$

$$73X_L - 352 = -60 \qquad 73X_C - 160 = 132$$

$$\underline{X_L = 4} \qquad \underline{X_C = 4}$$

10. $\lambda_1 = \frac{1}{3}(4\lambda_2 - 11)$ and $\lambda_2 = 3 + \frac{2}{3}\lambda_1$

$$3\lambda_1 + 11 = 4\left(3 + \frac{2}{3}\lambda_1\right) \quad \text{and} \quad 3\lambda_2 - \frac{2}{3}(4\lambda_2 - 11) = 9$$

$$9\lambda_1 - 8\lambda_1 = 3(12 - 11) \qquad 9\lambda_2 = 8\lambda_2 + 27 - 22$$

$$\underline{\lambda_1 = 3} \qquad \underline{\lambda_2 = 5}$$

12. $I_2 = \frac{1}{8}(18 - 6I_1)$ and $I_1 = \frac{1}{5}(22 - 4I_2)$

$18 - \frac{6}{5}(22 - 4I_2) = 8I_2$ and $5I_1 + \frac{1}{2}(18 - 6I_1) = 22$

$90 - 132 + 24I_2 = 40I_2$ $\qquad 5I_1 + 9 - 3I_1 = 22$

$\therefore \underline{I_2 = -\frac{21}{8}}$ $\qquad \therefore \underline{I_1 = \frac{13}{2}}$

14. $\omega = 2\pi - 8$ and $\pi = \frac{1}{3}(5 - 2\omega)$

$\frac{2}{3}(5 - 2\omega) - 8 = \omega$ and $2(2\pi - 8) + 3\pi = 5$

$10 - 4\omega - 24 = 3\omega$ $\qquad 4\pi + 3\pi = 21$

$-14 = 7\omega$ $\qquad 7\pi = 21$

$\therefore \underline{\omega = -2}$ $\qquad \therefore \underline{\pi = 3}$

16. $X_L = \frac{9}{4}X_C - 4$ and $X_C = \frac{1}{7}(6X_L - 2)$

$4X_L - \frac{9}{7}(6X_L - 2) = -16$ and $7X_C - 6\left(\frac{9}{4}X_C - 4\right) = -2$

$28X_L - 54X_L + 18 = -112$ $\qquad 28X_C - 54X_C + 96 = -8$

$-26X_L = -130$ $\qquad -26X_C = -104$

$\underline{X_L = 5}$ $\qquad \underline{X_C = 4}$

18. $I = \frac{2.6 - 0.8i}{0.6}$ and $i = \frac{0.5I - 7.0}{0.3}$

$0.6I + 0.8\left(\frac{0.5I - 7}{0.3}\right) = 2.6,$ and $7 - 0.5\left(\frac{2.6 - 0.8i}{0.6}\right) = -0.3i$

$0.58I = 6.38$ $\qquad 2.9 = -0.58i$

$\underline{I = 11}$ $\qquad \therefore \underline{i = -5}$

20. $M = 10\ 200 - 3L$ and $L = 3720 - 0.6M$

$0.6(3720 - 0.6M) = 2040 - 0.2M$

$-0.16M = -192$ $\qquad \therefore \underline{M = 1200}$

and $0.5L + 0.3(10\ 200 - 3L) = 1860$

$-0.4L = -1200$ $\qquad \therefore \underline{L = 3000}$

PROBLEMS 17-4

2. $I_1 = \frac{5 + I_2}{6}$ and $I_1 = 16 - 2I_2$

$\therefore 5 + I_2 = 96 - 12I_2$

$13I_2 = 91$ $\qquad \therefore \underline{I_2 = 7}$ and $\underline{I_1 = 2}$

4. $E = \frac{15 - 3e}{4} = \frac{36 - 11e}{2}$ and $e = \frac{15 - 4E}{3} = \frac{36 - 2E}{11}$

$30 - 6e = 144 - 44e$ 　　 $165 - 108 = 44E - 6E$

$38e = 114$ 　　 $57 = 38E$

$\therefore \underline{e = 3}$ 　　 $\therefore \underline{E = 1.5}$

6. $L_1 = \frac{6L_2 - 24}{5} = \frac{9L_2 - 22}{4}$ and $L_2 = \frac{4L_1 + 22}{9} = \frac{5L_1 + 24}{6}$

$24L_2 - 96 = 45L_2 - 110$ 　　 $24L_1 + 132 = 45L_1 + 216$

$-21L_2 = -14$ 　 $\therefore \underline{L_2 = \frac{2}{3}}$ 　　 $-21L_1 = 84$ 　 $\therefore L_1 = -4$

8. $M = 12Q = \frac{16 + 20Q}{3}$ and $Q = \frac{2M}{24} = \frac{-3M + 16}{20}$

$36Q = 16 + 20Q$ 　　 $40M = 72M - 384$

$\therefore Q = 1$ 　　 $-32M = -384$ 　 $\therefore \underline{M = 12}$

10. $I = \frac{6 + 5i}{2.1}$ and $i = \frac{2.1I - 6}{5}$ substitute for i and I and show that:

$\underline{i = 3}$ and $\underline{I = 10}$

PROBLEMS 17-5

2. $\frac{x}{3} - \frac{6y}{10} = 2$ Eq. (1)

$\frac{x}{3} - y = 4$ Eq. (2)

M 1×30: $10x - 18y = 60$ Eq. (3)

M 2×3: $x - 3y = 12$ Eq. (4)

$x = 12 + 3y$

Now substitute for x in Eq. (3)

$10(12 + 3y) - 18y = 60$

$120 + 30y - 18y = 60$

$12y = -60$

$\underline{y = -5}$ and $\underline{x = -3}$

4. $52E - 15e = 80$ Eq. (1)

$39E - 8e = 99$ Eq. (2)

1×8: $416E - 120e = 640$ Eq. (3)

2×15: $585E - 120e = 1485$ Eq. (4)

3-4: $-169E = -845$

$\therefore \underline{E = 5}$

Solving for e:

1×3: $156E - 45e = 240$ Eq. (5)

2×4: $156E - 32e = 396$ Eq. (6)

5-6: $-13e = -156$

$\therefore \underline{e = 12}$

6. $5I + 3i = -1$ Eq. (1)

$3I + 7i = 15$ Eq. (2)

1×3: $15I + 9i = -3$ Eq. (3)

2×5: $15I + 35i = 75$ Eq. (4)

3-4: $-26i = -78$

$\underline{i = 3}$

Solving for I:

1×7: $35I + 21i = -7$ Eq. (5)

2×3: $9I + 21i = 45$ Eq. (6)

5-6: $26I = -52$

$\therefore \underline{I = -2}$

8. $3(Z_1 + 2Z_2) - 18(Z_2 - 5) = 2(Z_1 + Z_2 + 1)$

and $Z_1 - 2 = 20 + 4Z_2 - 4Z_2 - 12$

$Z_1 - 14Z_2 = -88$ Eq. (1)

$Z_1 = 10$ Eq. (2) (This gives solution for Z_1)

1-2: $-14Z_2 = -98$

$\therefore Z_2 = 7$ $\therefore \underline{Z_1 = 10}$ and $\underline{Z_2 = 7}$

10. $320I - 248i = -15$ Eq. (1)

$-16I + 16i = 3$ Eq. (2)

2×20: $-320I + 320i = 60$ Eq. (3)

$320I - 248i = -15$ Eq. (1)

1+3: $+ 72i = 45$

$\therefore \underline{i = \frac{5}{8}}$

Substitute $i = \frac{5}{8}$ into Eq. (2) and solve for I.

$\therefore -16I + 10 = 3$

$-16I = -7$

$\therefore \underline{I = \frac{7}{16}}$

PROBLEMS 17-6

2. $\frac{12}{V} - \frac{9}{v} = 3$ Eq. (1)

$\frac{6}{V} + \frac{3}{v} = 4$ Eq. (2)

M 1×1: $\frac{12}{V} - \frac{9}{v} = 3$ Eq. (1)

M 2×2: $\frac{12}{V} + \frac{6}{v} = 8$ Eq. (3)

Eq. (3) - Eq. (1): $\frac{15}{v} = 5$ $\therefore \underline{v = 3}$

M 1×1: $\frac{12}{V} - \frac{9}{v} = 3$ Eq. (1)

M 2×3: $\frac{18}{V} + \frac{9}{v} = 12$ Eq. (4)

Eq. (1) + Eq. (4): $\frac{30}{V} = 15$ $\therefore \underline{V = 2}$

4. $\frac{6}{p} = \frac{1}{3}$ Eq. (1)

$\frac{-3}{p} + \frac{5}{q} = \frac{1}{4}$ Eq. (2)

Eq. (1) shows $\underline{P = 18}$

From Eq. (1), $P = 18$.

Substitute $P = 18$ into Eq. (2) and solve for q:

$\therefore -\frac{3}{18} + \frac{5}{q} = \frac{1}{4}$

$\therefore \underline{q = 12}$

6. $4 - 4b = 3a - 3 \equiv 3a + 4b = 7$ Eq. (1)

$10a - 195 = 14b - 35 \equiv 10a - 14b = 160$ Eq. (2)

1×10: $30a + 40b = 70$ Eq. (3)

2×3: $30a - 42b = 480$ Eq. (4)

3-4: $82b = -410$

$\underline{b = -5}$

To solve for a, substitute $b = -5$ in Eq. (1).

$\therefore 3a - 20 = 7$ $\therefore \underline{a = 9}$

8. $\lambda + 3\pi + 7 = 7\pi \equiv \lambda - 4\pi = -7$ Eq. (1)

$2\pi = 4\lambda \equiv -4\lambda + 2\pi = 0$ Eq. (2)

Use any method to show $\underline{\lambda = 1}$ and $\underline{\pi = 2}$

10. $\frac{1}{2\pi} + \frac{1}{2\lambda} = \frac{136}{70}$ Eq. (1)

$\frac{1}{2\pi} + \frac{1}{4\lambda} = \frac{108}{70}$ Eq. (2)

1-2: $\frac{1}{4\lambda} = \frac{28}{70}$

$\therefore \underline{\lambda = \frac{5}{8}}$

Substitute $\lambda = \frac{5}{8}$ into Eq. (1) and solve for π:

$\therefore \frac{1}{2\pi} + \frac{8}{10} = \frac{136}{70}$

$\therefore \underline{\pi = \frac{7}{16}}$

PROBLEMS 17-7

2. $5x + 2y = \alpha$ Eq. (1)

$2x - 7y = \beta$ Eq. (2)

M 1×2: $10x + 4y = 2\alpha$ Eq. (3)

M 2×5: $10x - 35y = 5\beta$ Eq. (4)

Eq. (3) - Eq. (4) $39y = 2\alpha - 5\beta$

$\therefore \underline{y = \frac{2\alpha - 5\beta}{39}}$

M 1×7: $35x + 14y = 7\alpha$ Eq. (5)

M 2×2: $4x - 14y = 2\beta$ Eq. (6)

Eq. (5) + Eq. (6): $39x = 7\alpha + 2\beta$

$\therefore \underline{x = \frac{7\alpha + 2\beta}{39}}$

4. $4L_1 + 3L_2 = C$ Eq. (1)

$3L_1 - 2L_2 = C$ Eq. (2)

1×2: $8L_1 + 6L_2 = 2C$ Eq. (3)

2×3: $9L_1 - 6L_2 = 3C$ Eq. (4)

3+4: $17L = 5C$

$\therefore \underline{L_1 = \frac{5C}{17}}$

Substitute $L_1 = \frac{5C}{17}$ into Eq. (1) and solve for L_2:

$\frac{20C}{17} + 3L_2 = C$

$3L_2 = \frac{(17 - 20)C}{17}$

$\therefore \underline{L_2 = \frac{-C}{17}}$

6. $15r + 9R = 3Z_1$ Eq. (3)

$15r + 35R = 5Z_2$ Eq. (4)

3-4: $-26R = 3Z_1 - 5Z_2$

$\therefore \underline{R = \frac{5Z_2 - 3Z_1}{26}}$

and

$35r + 21R = 7Z_1$ Eq. (5)

$9r + 21R = 3Z_2$ Eq. (6)

5-6: $26r = 7Z_1 - 3Z_2$

$\therefore \underline{r = \frac{7Z_1 - 3Z_2}{26}}$

8. $3R_L + 4R_P = 12R_T$ Eq. (1)

$2R_L - R_P = 4R_1$ Eq. (2)

1×2: $6R_L + 8R_P = 24R_T$ Eq. (3)

2×3: $6R_L - 3R_P = 12R_1$ Eq. (4)

3-4: $+11R_P = -12R_1 + 24R_T$

$$\therefore R_P = \frac{24R_T - 12R_1}{11}$$

$$R_P = 1\frac{1}{11}(2R_T - R_1)$$

$3R_L + 4R_P = 12R_T$ Eq. (1)

2×4: $8R_L - 4R_P = 16R_1$ Eq. (5)

1+5 $11R_L = 12R_T + 16R_1$

$$\therefore R_L = \frac{4}{11}(3R_T + 4R_1)$$

10. $Z_1 - Z_2 = 3Z_1 - 3Z_2 - 3X_C$

$\therefore 3X_C = 2Z_1 - 2Z_2$ Eq. (1)

Eq. (2) × 2: $0 = \frac{4}{5}Z_1 - 2Z_2$

$3X_C = \frac{6}{5}Z_1$ $\therefore Z_1 = \frac{5}{2}X_C$

Eq. (1) × $\frac{1}{5}$: $\frac{3}{5}X_C = \frac{2}{5}Z_1 - \frac{2}{5}Z_2$

From line 3: $0 = \frac{2}{5}Z_1 - \frac{5}{5}Z_2$ Eq. (2)

Substitute Eq. (2): $\frac{3}{5}X_C = \frac{3}{5}Z_2$

$\therefore Z_2 = X_C$

PROBLEMS 17-8

2. $I_1 - I_2 + I_3 = 6$ Eq. (1)

$I_1 + I_2 + I_3 = 10$ Eq. (2)

$I_1 + I_2 - I_3 = 0$ Eq. (3)

Add 1 and 3:

$$\begin{array}{r} I_1 - I_2 + I_3 = 6 \\ I_1 + I_2 - I_3 = 0 \\ \hline 2I_1 \qquad\qquad = 6 \end{array}$$

$$\therefore I_1 = 3$$

Subtract 1 from 2:

$$\begin{array}{r} I_1 + I_2 + I_3 = 10 \\ I_1 - I_2 + I_3 = 6 \\ \hline 2I_2 \qquad = 4 \end{array}$$

$$\therefore \underline{I_2 = 2}$$

Substitute for I_1 and I_2 in Eq. (3) and solve for I_3:

$$3 + 2 - I_3 = 0$$

$$\therefore \underline{I_3 = 5}$$

4. $a - 2b + c = 3$ Eq. (1)

$a + b + 2c = 1$ Eq. (2)

$2a - b + c = 2$ Eq. (3)

2+3: $3a + 3c = 3$ Eq. (4)

2×2: $2a + 2b + 4c = 2$

Eq. (1): $a - 2b + c = 3$

(2×2) + Eq. (1) = $3a + 5c = 5$ Eq. (5)

$3a + 3c = 3$ Eq. (4)

5-4: $2c = 2$ $\quad \therefore \underline{c = 1}$

Substitute for $c = 1$ in Eq. (4):

$3a + 3 = 3 \quad \therefore \underline{a = 0}$

Substitute for a and c in Eq. (1):

$0 - 2b + 1 = 3 \quad \therefore \underline{b = -1}$

Answer: $\underline{a = 0}$, $\underline{b = -1}$, $\underline{c = 1}$

6. $\frac{1}{a} - \frac{1}{b} - \frac{1}{c} = 1$ Eq. (1)

$-\frac{1}{a} + \frac{1}{b} - \frac{1}{c} = 1$ Eq. (2)

$-\frac{1}{a} - \frac{1}{b} + \frac{1}{c} = 1$ Eq. (3)

Add Eq. (1) and Eq. (2): $-\frac{2}{c} = 2$ $\quad \therefore c = -1$

Add Eq. (1) and Eq. (3): $-\frac{2}{b} = 2$ $\quad \therefore b = -1$

Substitute for b and c in Eq. 1:

$$\frac{1}{a} + 1 + 1 = 1 \qquad \therefore \underline{a = -1}$$

8. $E_1 - E_2 - E_3 = \alpha$ Eq. (1)

$-E_1 - E_2 + E_3 = \beta$ Eq. (2)

$-E_1 + E_2 - E_3 = \gamma$ Eq. (3)

Add Eq. (1) and Eq. (2):

$$-2E_2 = \alpha + \beta$$

$$\therefore \underline{E_2 = -\frac{\alpha + \beta}{2}}$$

Add Eq. (2) to Eq. (3):

$$-2E_1 = \beta + \gamma$$

$$\therefore \underline{E_1 = -\frac{\beta + \gamma}{2}}$$

Add Eq. (1) to Eq. (3) to show:

$$\therefore \underline{E_3 = -\frac{\alpha + \gamma}{2}}$$

10. $a - c = -5$ Eq. (1)

$7b - 3c = -1$ Eq. (2)

$2b - a - c = -9$ Eq. (3)

Data for solution:

$a = c - 5$, $b = \frac{3c - 1}{7}$,

$c = \frac{7b + 1}{3}$.

Substitute for a and b in Eq. (3): $2\left(\frac{3c - 1}{7}\right) - (c - 5) - c = -9$

$$\frac{6c - 2}{7} - \frac{14c}{7} = -14$$

$$-8c = -96$$

$$\therefore \underline{c = 12}$$

Substitute for c in Eq. (1):

$a - 12 = -5$ $\qquad \therefore \underline{a = 7}$

Substitute for c in Eq. (2):

$7b - 36 = -1$ $\qquad \therefore \underline{b = 5}$

PROBLEMS 17-9

2. $a + b = 16$

$\underline{a - b = \ \ 4}$

$2a \quad\; = 20$ $\qquad \therefore \underline{a = 10}$ and $\underline{b = 6}$

4. $\alpha + \beta = 90^\circ$

$\underline{\alpha - \beta = 10^\circ}$

$2\alpha \quad = 100^\circ$

$\underline{\alpha = 50^\circ}$ $\qquad \therefore \underline{\beta = 40^\circ}$

6. $a + b + c = 180^\circ$

$a = \frac{b^\circ}{3}$ and $c = b + 5^\circ$

$\frac{b}{3} + b + b + 5^\circ = 180^\circ$

$\frac{7b^\circ}{3} = 175^\circ$

$\therefore \underline{b = 75^\circ}, \underline{a = 25^\circ}, \underline{c = 80^\circ}$

8. $\frac{25}{A} = \frac{25}{B} + 1$ Eq. (1)

$\frac{25}{2A} = \frac{25}{B} - 1.5$ Eq. (2)

$\frac{25}{2A} + 1.5 = \frac{25}{A} - 1$

$\frac{25}{2} + 1.5A = 25 - A$

$\underline{A = 5 \text{ km/h}}$ and $\underline{B = 6.25 \text{ km/h}}$

10. $g = \frac{v}{t}$, $\therefore s = \frac{1}{2}\frac{v}{t}t^2$

$\therefore \underline{v = 2\frac{s}{t}}$

12. $2.7 \text{ kW} = 180I$ $\therefore I = \underline{15 \text{ A}}$

and $R = \frac{180}{15} = \underline{12\ \Omega}$

14. $\frac{2s}{t} - u = v = u + at$

$t = \frac{2s}{u + v}$, $v = u + \frac{2as}{u + v}$

$v^2 + vu = u^2 + vu + 2as$

$\therefore \underline{v^2 = u^2 + 2as}$

16. $\frac{2\pi fL}{Q} = 2D_L fL$

$D_L = \frac{2\pi fL}{Q2fL}$

$\therefore D_L = \frac{\pi}{Q}$

18. $IR = I_1R + I_1R_1$

$I_1R_1 = R(I - I_1)$ $\therefore \underline{R = \frac{I_1R_1}{I - I_1}}$

20. Given $P = IV$ and $V = IR$ $\therefore I = \frac{V}{R}$

Substitute for V: $P = (I)(IR)$ $\therefore P = I^2R$ or $I^2 = \frac{P}{R}$

$\therefore$ given $H = 0.24I^2Rt$ J (units)

Substitute $\frac{P}{R}$ for I^2 $H = 0.24\frac{P}{\cancel{R}}\cancel{R}t$ $\therefore H = 0.24Pt$ J

22. $\frac{Q_a}{t}R_a = \frac{Q_b}{t}R_b$

$\frac{Q_a}{Q_b}R_a = R_b$ and $\frac{Q_a}{Q_b} = \frac{C_a}{C_b}$

$\therefore \underline{R_aC_a = R_bC_b}$

24. $\frac{V_p}{R} = \frac{\mu V_g}{R + R_p}$

$\therefore V_pR + V_pR_p = \mu V_gR$

$R(V_p - \mu V_g) = -V_pR_p$

or: $R(\mu V_g - V_p) = V_pR_p$

$\therefore R = \frac{V_pR_p}{\mu V_g - V_p}$

$I_pR + I_pR_p = \mu V_g$

$\therefore R = \frac{\mu V_g - I_pR_p}{I_p}$

$\therefore R = \frac{250 \times 10^3}{12.5} - 10 \times 10^3$

$R = 10\ \text{k}\Omega$

$V_p = I_pR = 12.5 \times 10^{-3} \times 10 \times 10^3$

$\therefore \underline{V_p = 125\ \text{V}}$

26. See text

28. $I_t = I_1 + I_2 + I_3$

$\frac{1}{R_t} = \frac{1}{R_1} + \frac{1}{R_2} + \frac{1}{R_3}$

$I_1R_1 = I_2R_2 = I_3R_3 = I_tR_t$

Since $I_tR_t = I_1R_1$, solve for I_1

$I_1 = \frac{I_t}{\not{R_t}} \times \frac{\not{R_1}R_2R_3}{R_1R_2 + R_1R_3 + R_2R_3}$

$\therefore \underline{I_1} = I_t\left(\frac{R_2R_3}{R_1R_2 + R_1R_3 + R_2R_3}\right)$

$\underline{\text{and } I_2} = I_t\left(\frac{R_1R_3}{R_1R_2 + R_1R_3 + R_2R_3}\right)$

$\therefore \underline{I_3} = I_t\left(\frac{R_1R_2}{R_1R_2 + R_2R_3 + R_1R_3}\right)$

PROBLEMS 18-1

2. $\begin{vmatrix} 2 & 6 \\ 15 & -9 \end{vmatrix} \begin{matrix} -90 \\ -18 \end{matrix} = \underline{-108}$

4. $\begin{vmatrix} -2 & 3 \\ 7 & 15 \end{vmatrix} \begin{matrix} -21 \\ -30 \end{matrix} = -51$

6. $\begin{vmatrix} 3 & -2 \\ 1 & 2 \end{vmatrix} \begin{matrix} 2 \\ 6 \end{matrix} = \underline{8}$

8. $\begin{vmatrix} -3 & -7 \\ -4 & -2 \end{vmatrix} \begin{matrix} -28 \\ 6 \end{matrix} = -22$

10. $\begin{vmatrix} -0.06 & 0.02 \\ 0.05 & -1.6 \end{vmatrix} \begin{matrix} -0.001 \\ 0.096 \end{matrix} = \underline{0.095}$

12. $\begin{vmatrix} a & b \\ x & y \end{vmatrix} \begin{matrix} -bx \\ ay \end{matrix} = \underline{ay - bx}$

14. $\begin{vmatrix} a & x \\ b & y \end{vmatrix} = ay - bx$

16. $\begin{vmatrix} y & x \\ b & a \end{vmatrix} \begin{matrix} -bx \\ ay \end{matrix} = \underline{ay - bx}$

Note: The learner should explain why the answers to Problems 12, 14, and 16 have identical solutions.

PROBLEMS 18-2

2.
$$x = \frac{\begin{vmatrix} 20 & 1 \\ 25 & 3 \end{vmatrix}}{\begin{vmatrix} 3 & 1 \\ 2 & 3 \end{vmatrix}} = \frac{60 - 25}{9 - 2} = \underline{5} \text{ and } y = \frac{\begin{vmatrix} 3 & 20 \\ 2 & 25 \end{vmatrix}}{7} = \frac{75 - 40}{7} = \underline{5}$$

4.
$$R_1 = \frac{\begin{vmatrix} 29 & 3 \\ -7 & -3 \end{vmatrix}}{\begin{vmatrix} 1 & 3 \\ 1 & -3 \end{vmatrix}} = \underline{+11} \text{ and } R_2 = \frac{\begin{vmatrix} 1 & 29 \\ 1 & -7 \end{vmatrix}}{-6} = \underline{+6}$$

6.
$$V_g = \frac{\begin{vmatrix} 1 & 3 \\ -2 & 1 \end{vmatrix}}{\begin{vmatrix} 2 & 3 \\ 1 & 1 \end{vmatrix}} = \underline{-7} \text{ and } V = \frac{\begin{vmatrix} 2 & 1 \\ 1 & -2 \end{vmatrix}}{-1} = \underline{5}$$

8.
$$X_C = \frac{\begin{vmatrix} 2.9 & 3 \\ 17 & 30 \end{vmatrix}}{\begin{vmatrix} 4 & 3 \\ 8 & 30 \end{vmatrix}} = \underline{0.375}, \text{ or } \left(\frac{3}{8}\right) \text{ and } X_L = \frac{\begin{vmatrix} 4 & 2.9 \\ 8 & 17 \end{vmatrix}}{96} = \underline{\frac{44.8}{96}}, \text{ or } \left(\frac{7}{15}\right)$$

10.
$$Z_1 = \frac{\begin{vmatrix} 9300 & 1 \\ 192 & -0.06 \end{vmatrix}}{\begin{vmatrix} 1 & 1 \\ 0.04 & -0.06 \end{vmatrix}} = \underline{7500} \text{ and } Z_2 = \frac{\begin{vmatrix} 1 & 9300 \\ 0.04 & 192 \end{vmatrix}}{-0.1} = \underline{1800}$$

PROBLEMS 18-3

2.
$$\left|\begin{array}{ccc|cc} 2 & 5 & 1 & 2 & 5 \\ 7 & 18 & 4 & 7 & 18 \\ -1 & -20 & -5 & -1 & -20 \end{array}\right| \quad \frac{+18 + 160 + 175}{-180 - 20 - 140} = \underline{13}$$

*See Prob. 8, R_2 solution.

4.
$$\left|\begin{array}{ccc|cc} -3 & -2 & 3 & -3 & -2 \\ 0 & -7 & 2 & 0 & -7 \\ 0 & 7 & -4 & 0 & 7 \end{array}\right| \quad \frac{0 + 42 + 0}{-84 + 0 + 0} = \underline{-42}$$

6.
$$\left|\begin{array}{ccc|cc} 3 & 8 & 3.2 & 3 & 8 \\ 12 & 20 & 16.5 & 12 & 20 \\ -16 & -12 & -7.8 & -16 & -12 \end{array}\right| \quad \frac{1024 + 594 + 748.8}{-468 - 2112 - 460.8} = \underline{-674}$$

8.

$$\text{Denominator} = \left|\begin{array}{rrr|rr} 1 & 1 & 1 & 1 & 1 \\ 5 & -2 & 6 & 5 & -2 \\ -2 & 3 & -3 & -2 & 3 \end{array}\right| \begin{array}{l} -4 - 18 + 15 \\ \\ 6 - 12 + 15 \end{array} = 2$$

*Before solving for R_1 see solution for Prob. 2; this is the numerator for R_2.

$$\therefore R_2 = \frac{-6}{2} = \underline{-3}$$

Solve for R_1 and R_3:

$$R_1 = \frac{\left|\begin{array}{rrr|rr} 3 & 1 & 1 & 3 & 1 \\ 40 & -2 & 6 & 40 & -2 \\ -25 & 3 & -3 & -25 & 3 \end{array}\right| \begin{array}{l} -50 - 54 + 120 \\ \\ 18 - 150 + 120 \end{array}}{2} = \frac{4}{2} = \underline{2}$$

$$R_3 = \frac{\left|\begin{array}{rrr|rr} 1 & 1 & 3 & 1 & 1 \\ 5 & -2 & 40 & 5 & -2 \\ -2 & 3 & -25 & -2 & 3 \end{array}\right| \begin{array}{l} -12 - 120 + 125 \\ \\ 50 - 80 + 45 \end{array}}{2} = \frac{8}{2} = \underline{4}$$

Answers: $R_1 = \underline{2}$, $R_2 = \underline{-3}$, and $R_3 = \underline{4}$

10.

$$\text{Denominator} = \left|\begin{array}{rrr|rr} 3 & 5 & -2 & 3 & 5 \\ -4 & 1 & 1 & -4 & 1 \\ 2 & 3 & -7 & 2 & 3 \end{array}\right| = -132$$

$$\therefore r = \frac{\left|\begin{array}{rrr|rr} -3 & 5 & -2 & -3 & 5 \\ 0 & 1 & 1 & 0 & 1 \\ -42 & 3 & -7 & -42 & 3 \end{array}\right|}{-132} = \frac{-264}{-132} = \underline{2}$$

$$\text{and } p = \frac{\left|\begin{array}{rrr|rr} 3 & -3 & -2 & 3 & -3 \\ -4 & 0 & 1 & -4 & 0 \\ 2 & -42 & -7 & 2 & -42 \end{array}\right|}{-132} = \frac{-132}{-132} = \underline{1}$$

$$\text{and } q = \frac{\left|\begin{array}{rrr|rr} 3 & 5 & -3 & 3 & 5 \\ -4 & 1 & 0 & -4 & 1 \\ 2 & 3 & -42 & 2 & 3 \end{array}\right|}{-132} = \frac{-924}{-132} = \underline{7}$$

12.

$$\text{Denominator} = \begin{vmatrix} 12 & 20 & 10 \\ 3 & 8 & -6 \\ -16 & -12 & 20 \end{vmatrix}\begin{matrix} 12 & 20 \\ 3 & 8 \\ -16 & -12 \end{matrix}\Bigg| = \underline{2696}$$

$$I_1 = \frac{\begin{vmatrix} 16.5 & 20 & 10 \\ 3.2 & 8 & -6 \\ -7.8 & -12 & 20 \end{vmatrix}\begin{matrix} 16.5 & 20 \\ 3.2 & 8 \\ -7.8 & -12 \end{matrix}\Bigg|}{2696} = \frac{1348}{2696} = \underline{0.5 \text{ A}}$$

$$I_2 = \frac{\begin{vmatrix} 12 & 16.5 & 10 \\ 3 & 3.2 & -6 \\ -16 & -7.8 & 20 \end{vmatrix}\begin{matrix} 12 & 16.5 \\ 3 & 3.2 \\ -16 & -7.8 \end{matrix}\Bigg|}{2696} = \frac{1078.4}{2696} = \underline{0.4 \text{ A}}$$

Using the same method as in the I_1 and I_2 solutions, solve for I_3: $I_3 = \frac{674}{2696} = \underline{0.25 \text{ A}}$

PROBLEMS 18-4

2. $\begin{vmatrix} 1 & -1 \\ -3 & -1 \end{vmatrix} \begin{matrix} -3 \\ -1 \end{matrix} = \underline{-4}$

4. $\begin{vmatrix} -3 & -7 \\ -8 & 2 \end{vmatrix} \begin{matrix} -56 \\ -6 \end{matrix} = \underline{-62}$

6. $\begin{vmatrix} 0 & 5 \\ 2 & -20 \end{vmatrix} \begin{matrix} -10 \\ 0 \end{matrix} = \underline{-10}$

8. $\begin{vmatrix} 3 & -3 & 2 \\ 2 & 2 & -1 \\ 0 & -5 & 1 \end{vmatrix}\begin{matrix} 3 & -3 \\ 2 & 2 \\ 0 & -5 \end{matrix}\Bigg| \begin{matrix} 0 - 15 + 6 \\ \\ 6 + 0 - 20 \end{matrix} = \underline{-23}$

PROBLEMS 18-5

2. $-1\begin{vmatrix} 7 & -1 \\ 2 & 5 \end{vmatrix} - 4\begin{vmatrix} 3 & -1 \\ 2 & 5 \end{vmatrix} + 3\begin{vmatrix} 3 & -1 \\ 7 & -1 \end{vmatrix}$

$= -1(35 + 2) - 4(15 + 2) + 3(-3 + 7) = \underline{-93}$

4. $-35\begin{vmatrix} 0 & -12 \\ -6 & 3 \end{vmatrix} - 96\begin{vmatrix} 5 & 1 \\ -6 & 3 \end{vmatrix} + 18\begin{vmatrix} 5 & 1 \\ 0 & -12 \end{vmatrix}$

$= -35(0 - 72) - 96(15 + 6) + 18(-60 - 0) = \underline{-576}$

6. $-8\begin{vmatrix} 4 & 10 \\ 16 & 2 \end{vmatrix} - 16\begin{vmatrix} 2 & 10 \\ 0 & 2 \end{vmatrix} - 13\begin{vmatrix} 2 & 4 \\ 0 & 16 \end{vmatrix}$

$= +8(8 - 160) - 16(4) + 13(32) = \underline{-864}$

8. $$11\begin{vmatrix}16 & 12 & -10 & -2\\ 2 & 2 & 3 & -9\\ 0 & 2 & 15 & 4\\ -4 & 10 & -8 & 0\end{vmatrix} - 5\begin{vmatrix}2 & 16 & 12 & -2\\ 5 & 2 & 2 & -9\\ 5 & 0 & 2 & 4\\ 0 & -4 & 10 & 0\end{vmatrix} + 4\begin{vmatrix}2 & 16 & 12 & -10\\ 5 & 2 & 2 & 3\\ 5 & 0 & 2 & 15\\ 0 & -4 & 10 & -8\end{vmatrix}$$

$a \qquad b \qquad c$

a

$$11\left[16\begin{vmatrix}2 & 3 & -9\\ 2 & 15 & 4\\ 10 & -8 & 0\end{vmatrix} - 2\begin{vmatrix}12 & -10 & -2\\ 2 & 15 & 4\\ 10 & -8 & 0\end{vmatrix} + 4\begin{vmatrix}12 & -10 & -2\\ 2 & 3 & -9\\ 2 & 15 & 4\end{vmatrix}\right]$$

$x \qquad y \qquad z$

x

$$176\left[2\begin{vmatrix}15 & 4\\ -8 & 0\end{vmatrix} - 2\begin{vmatrix}3 & -9\\ -8 & 0\end{vmatrix} + 10\begin{vmatrix}3 & -9\\ 15 & 4\end{vmatrix}\right] = 176(1678) \qquad = 295\ 328$$

y

$$-22\left[12\begin{vmatrix}15 & 4\\ -8 & 0\end{vmatrix} - 2\begin{vmatrix}-10 & -2\\ -8 & 0\end{vmatrix} + 10\begin{vmatrix}3 & -9\\ 15 & 4\end{vmatrix}\right] = -22(316) \qquad = -6952$$

z

$$44\left[12\begin{vmatrix}3 & -9\\ 15 & 4\end{vmatrix} - 2\begin{vmatrix}-10 & -2\\ 15 & 4\end{vmatrix} + 2\begin{vmatrix}-10 & -2\\ 3 & -9\end{vmatrix}\right] = 44(1976) \qquad = \underline{86\ 944}$$

$$x + y + z = a = 375\ 320$$

Solve for b: $$a + b = 310\ 560$$

$$b = -5\begin{vmatrix}2 & 16 & 12 & -2\\ 5 & 2 & 2 & -9\\ 5 & 0 & 2 & 4\\ 0 & -4 & 10 & 0\end{vmatrix} = -64\ 760 \text{ and } c = 4\begin{vmatrix}2 & 16 & 12 & -10\\ 5 & 2 & 2 & 3\\ 5 & 0 & 2 & 15\\ 0 & 4 & 10 & -8\end{vmatrix} = 46\ 336$$

$$356\ 896$$

$\therefore \underline{a + b + c = 356\ 896}$

10. $$\begin{vmatrix}2 & 1 & -1\\ 3 & -2 & 2\\ 4 & -3 & -1\end{vmatrix} = \text{denominator} = 28$$

$$\theta = \frac{\begin{vmatrix}3 & 1 & -1\\ 8 & -2 & 2\\ -13 & -3 & -1\end{vmatrix}}{28} = \underline{2} \qquad \phi = \frac{\begin{vmatrix}2 & 3 & -1\\ 3 & 8 & 2\\ 4 & -13 & -1\end{vmatrix}}{28} = \frac{140}{28} = \underline{5}$$

$\lambda = 2\theta + \phi - 3 \qquad \therefore \lambda = 4 + 5 - 3 \qquad \therefore \underline{\lambda = 6},\ \underline{\theta = 2},\ \underline{\phi = 5}$

12. $$\text{Denominator} = \begin{vmatrix}2 & 3 & 2\\ -8 & 2 & -10\\ -3 & -7 & 4\end{vmatrix} = 2\begin{vmatrix}2 & -10\\ -7 & 4\end{vmatrix} + 8\begin{vmatrix}3 & 2\\ -7 & 4\end{vmatrix} - 3\begin{vmatrix}3 & 2\\ 2 & -10\end{vmatrix}$$

$$= 2(8 - 70) + 8(12 + 14) - 3(-30 - 4)$$

$$= -124 + 208 + 102$$

$$= 186$$

$I_1 = \dfrac{-1116}{186} = \underline{-6} \qquad I_2 = \underline{-2} \qquad I_3 = \underline{-4}$

14. $$\begin{vmatrix}2 & 4 & 10\\ -8 & -16 & 5\\ 0 & 16 & -20\end{vmatrix} = 2\begin{vmatrix}-16 & 5\\ 16 & 20\end{vmatrix} + 8\begin{vmatrix}4 & 10\\ 16 & -20\end{vmatrix} = \text{Denominator} = -1440$$

Problem 18-5 Number 14, continued.

$$x = \frac{\begin{vmatrix} 10 & 4 & 10 \\ -13 & -16 & 5 \\ 2 & 16 & -20 \end{vmatrix}}{-1440} = \underline{0.25}, \quad \underline{y = 0.875}, \quad \underline{z = 0.60}$$

16. Denominator:

$$\begin{vmatrix} 2 & 3 & -1.5 & 2.2 & 3 \\ -3 & 1 & 2 & -5.2 & 2 \\ 3 & -2 & -4.2 & -8 & -1.6 \\ 2 & -3 & 2 & 4.6 & -2.4 \\ 5 & 4 & -4.2 & -1.4 & 2.6 \end{vmatrix}$$

Numerator I_1:

$$\begin{vmatrix} -9.6 & 3 & -1.5 & 2.2 & 3 \\ 8 & 1 & 2 & -5.2 & 2 \\ 2.7 & -2 & -4.2 & -8 & -1.6 \\ 6.1 & -3 & 2 & 4.6 & -2.4 \\ -3.7 & 4 & -4.2 & -1.4 & 2.6 \end{vmatrix}$$

Numerator I_2:

$$\begin{vmatrix} 2 & (-)-9.6 & -1.5 & 2.2 & 3 \\ -3 & (+)8 & 2 & -5.2 & 2 \\ 3 & (-)2.7 & -4.2 & -8 & -1.6 \\ 2 & (+)6.1 & 2 & 4.6 & -2.4 \\ 5 & (-)3.7 & -4.2 & -1.4 & 2.6 \end{vmatrix} \Rightarrow (-)-9.6 \begin{vmatrix} -3 & 2 & -5.2 & 2 \\ 3 & -4.2 & -8 & -1.6 \\ 2 & 2 & 4.6 & -2.4 \\ 5 & -4.2 & -1.4 & 2.6 \end{vmatrix}$$

$\therefore \underline{I_1 = 2.5}$ $\quad \underline{I_2 = 3.1}$ $\quad \underline{I_3 = 5}$ $\quad \underline{I_4 = -2}$ $\quad \underline{I_5 = -4}$

*Student exercise, to prove statements

PROBLEMS 19-1

2. $\frac{38.4}{476.24} = \underline{80.6\text{ mA}}$

4. From Prob. 3, $r_{int} = 0.169\ \Omega$

$P = I^2R$

$\therefore \underline{P = 59\text{ mW}}$

PROBLEMS 19-2

2. (a) $r_{in} = \frac{1.4 - 1.099}{1.5} = \underline{0.2\ \Omega}$

(b) $P = I^2R = (1.5)^2(0.2) = \underline{450\text{ mW}}$

4. $0.33 \times 4 = \underline{1.32\text{ V}}$

6. No answer provided.

8. From Prob. 7, $R_L = 9.02\ \Omega$

$\frac{V^2}{R} = P = \underline{283\text{ mW}}$

10. $I = \frac{1.5}{33} = \underline{45.4\text{ mA}}$

12. (a) $V = 3.2$ V, $r_{\text{int}} = 0.067\ \Omega$ (total), $R_L = 0.85\ \Omega$

$$I_T = \frac{3.2}{0.067 + 0.85} = 3.49\text{ A} \qquad \therefore V_{R_L} = 2.966\text{ V}$$

$$P_{R_L} = \frac{(2.966)^2}{0.85} = \underline{10.4\text{ W}}$$

(b) Each battery draws $\frac{1}{6}I_T = \frac{3.49}{6} = \underline{582\text{ mA}}$

14. Battery emf = 2.1 V per cell, and 48 cells = 100.8 V

Battery r_{in} = 0.02 Ω per cell, and 48 cells = 0.96 Ω

Maximum current is to be 10 A

(a) $10 = \frac{115 - 100.8}{r_s + 0.96} \qquad \therefore r_s = \frac{115 - 100.8 - 9.6}{10} = \underline{0.46\ \Omega}$

(b) Power $= I^2(r_s + r_{in}) = 100(0.46 + 0.96) = \underline{142\text{ W}}$

(c) $P_{r_s} = I^2 r_s = 100(0.46) = \underline{46\text{ W}}$

(d) $I_{\text{short cct}} = \frac{V_{\text{batt}}}{r_{in}} = \frac{100.8}{0.96} = \underline{105\text{ A}}$

16. Series cct. $= 10\text{ V} = 3(1 + 10r)$

$$10\text{ V} = 3 + 30r \qquad \text{Eq. (1)}$$

Parallel cct. $V = 6\left(0.1 + \frac{r}{3}\right)$

$$V = 0.6 + 2r \qquad \text{Eq. (2)}$$

Solve Eq. (1) and Eq. (2) by determinants: $\underline{V = 1.2\text{ V}}$ and $\underline{r = 0.3\ \Omega}$

18. $12V = (I)(R + 12r) \qquad 24 = I(2.4 + 12r)$

$$V = (I)\left(R + \frac{r}{5}\right) \qquad 2 = I\left(0.24 + \frac{r}{5}\right)$$

$$I = \frac{10}{1 + 5r} \quad \text{Eq. (1)}, \qquad I = \frac{10}{1.2 + r} \quad \text{Eq. (2)}$$

Eq. (1) = Eq. (2)

$$\therefore 10(1.2 + r) = 10(1 + 5r)$$

$$40r = 2 \qquad \therefore \underline{r = 0.05\ \Omega}$$

$$24 = I[2.4 + 12(0.05)] \therefore \underline{I = 8\text{ A}}$$

20. $\underline{r_i = 0.02\ \Omega}$, $\underline{V_o = 1.6\text{ V}}$

PROBLEMS 20-1

2. θ^6 4. ε^9 6. θ^{5x} 8. w^{2a} 10. $\alpha^{2.6}$ 12. $e^{\pi-1}$

14. $\psi^{\beta+\gamma}$ 16. x^{15} 18. $I^6R^3t^3$ 20. a^{4x} 22. $-x^{3a}b^{3y}$ 24. $R_1^3R_2^3R_3^{-3}$

26. $\dfrac{Z_1^4}{Z_3^2Z_4^2}$ 28. $\dfrac{\pi^4D^8}{256}$ 30. $a^{4\gamma(3\pi\ -\ 5\lambda)}$ 32. $\dfrac{1}{a^4y^3}$

34. $\dfrac{25}{X_L^4X_C^2} = \left(\dfrac{5}{X_L^2X_C}\right)^2$ 36. $\dfrac{1}{(\pi R^2)^{2L}}$ 38. $\left(\dfrac{Z_4}{Z_1Z_2}\right)^3$ 40. $8\alpha^3\beta^2\gamma^2$

PROBLEMS 20-2

2. $\sqrt[3]{-64} = -4$

4. $-\sqrt[6]{-64} = -\sqrt{\sqrt[3]{64}} = -\sqrt{4} = -2$

6. $\pm\sqrt{L_1^4L_2^4} = \pm L_1^2L_2^2$

8. $\pm X^3C^3$

10. $\left(\dfrac{r^{12}R^8}{16V^4}\right)^{\frac{3}{4}} = \left(\dfrac{r^{12}R^8}{2^4V^4}\right)^{\frac{3}{4}} = \pm\ \dfrac{r^9R^6}{2^3V^3} = \pm\ \dfrac{r^9R^6}{8V^3}$

12. $\sqrt{15}$ 14. $\sqrt[3]{36}$ 16. $\sqrt[3]{x^2} \times \sqrt{y^3} = x^{\frac{2}{3}}y^{\frac{3}{2}} = \sqrt[6]{x^4y^9}$

18. $M^{\frac{2}{3}}$ 20. $a^{\frac{2}{3}}b^{\frac{4}{3}}c^2$ 22. $\alpha\beta^{\frac{2}{5}}$ 24. $4L\omega^{\frac{3}{2}}$

26. $-5\alpha^2 32^{\frac{1}{5}}\alpha^{\frac{3}{5}}\beta^{\frac{7}{5}} = \underline{-10\alpha^{\frac{13}{5}}\beta^{\frac{7}{5}}}$

PROBLEMS 20-3

2. $\pm\ 4\sqrt{2}$ 4. $\pm\ 2\sqrt{14}$ 6. $\pm\ 2\sqrt{6}$ 8. $\pm\ 2\sqrt{7}$ 10. $\pm\ 3x^2\sqrt{3}$ 12. $\pm\ 3A\sqrt{11AD}$

14. $\pm\ 3\pi\left(\sqrt{9r^2z^4\pi^2} \times \sqrt{8rz\pi}\right)$

$\pm\ 3\pi\left(\sqrt{9rz^2\pi}\right) \times (\sqrt{2rz\pi}) \times \left(\sqrt{4\pi rz^2}\right)$

$\pm\ 3\pi(6rz^2\pi \times \sqrt{2rz\pi})$

$\underline{\pm\ 18r\pi^2z^2\sqrt{2r\pi z}}$

16. $\pm\ 7x\sqrt{49y^2z^2D^2} \times \sqrt{3xzD} = \pm\ 49xyzD\sqrt{3xzD}$

18. $\pm\ X_LZ_1^2(\sqrt{81} \times \sqrt{7}) = \pm\ 72X_LZ_1^2\sqrt{7}$

20. $\pm 5\theta^3\lambda^3(\sqrt{17\lambda\theta} \times \sqrt{17\lambda\theta})$

$= \pm 5\theta^3\lambda^3(17\sqrt{\lambda\theta})$

$= \pm 85\theta^3\lambda^3\sqrt{\lambda\theta}$

PROBLEMS 20-4 $\left(\text{Introductory solutions, where } \dfrac{\sqrt{7}}{\sqrt{7}} = 1\right)$

2. $\sqrt{\dfrac{1}{13}} = \sqrt{\dfrac{1}{13}} \times \sqrt{\dfrac{13}{13}} = \dfrac{\sqrt{13}}{13}$

4. $\sqrt{\dfrac{4}{5}} \times \sqrt{\dfrac{5}{5}} = \dfrac{\sqrt{20}}{5} = \dfrac{2\sqrt{5}}{5}$

6. $\dfrac{\sqrt{7} \times \sqrt{15}}{15} = \dfrac{\sqrt{105}}{15}$

8. $\dfrac{9\sqrt{3}}{3} = \pm 3\sqrt{3}$

10. $\dfrac{21\sqrt{35}\sqrt{7}}{7} = \underline{3\sqrt{245}}$ or $\dfrac{21\sqrt{7}\sqrt{5}\sqrt{7}}{7} = \underline{21\sqrt{5}}$

12. $\dfrac{\sqrt{QR}}{R}$

14. $\dfrac{\pi\sqrt{X_L}\sqrt{2\pi f L}}{2\pi f L} = \dfrac{\sqrt{2\pi f L X_L}}{2 f L}$

16. $\dfrac{2F\sqrt{f_0}\sqrt{F}}{f_0 F} = \dfrac{2\sqrt{f_0 F}}{f_0}$

18. $\dfrac{\sqrt{E - e}\sqrt{E + e}}{E + e} = \dfrac{\sqrt{E^2 - e^2}}{E + e}$

20. $\sqrt{R^2 + \dfrac{R^2}{9}} = \sqrt{\dfrac{9R^2 + R^2}{9}} = \dfrac{\sqrt{10R^2}}{\sqrt{9}} = \dfrac{R\sqrt{10}}{3}$

PROBLEMS 20-5

2. $5\sqrt{3} + 2\sqrt{3 \times 4} = 5\sqrt{3} + (2)(2)\sqrt{3} = 9\sqrt{3}$

4. $\sqrt{7 \times 12} - \sqrt{7 \times 4} = 11\sqrt{7} - 2\sqrt{7} = 9\sqrt{7}$

6. $\alpha\sqrt{2} + \beta\sqrt{8} - \gamma\sqrt{50} = \alpha\sqrt{2} + 2\beta\sqrt{2} - 5\gamma\sqrt{2}$

$= \underline{\sqrt{2}(\alpha + 2\beta - 5\gamma)}$

8. $3\sqrt{\dfrac{1}{3}} = \pm\sqrt{\dfrac{9}{3}} = \underline{\pm\sqrt{3}}$

10. $18\sqrt{3} + 20\sqrt{2}$

12. $\dfrac{R_1}{3} + \dfrac{R_1\sqrt{16}\sqrt{3}}{3} = \dfrac{R_1 + R_1\sqrt{48}}{3} = \dfrac{R_1}{3}(1 + 4\sqrt{3})$

14. $\dfrac{\sqrt{\varepsilon + \eta}}{\sqrt{\varepsilon - \eta}} + \dfrac{\sqrt{\varepsilon - \eta}}{\sqrt{\varepsilon + \eta}} = \dfrac{(\sqrt{\varepsilon + \eta})^2 + (\sqrt{\varepsilon - \eta})^2}{(\sqrt{\varepsilon + \eta})(\sqrt{\varepsilon - \eta})} = \dfrac{\varepsilon + \eta + \varepsilon - \eta}{\sqrt{\varepsilon^2 - \eta^2}}$

Rationalizing denominator: $\therefore$ answer $= \dfrac{2\varepsilon\sqrt{\varepsilon^2 - \eta^2}}{\varepsilon^2 - \eta^2}$

16. $\dfrac{\sqrt{7R^2}\sqrt{16V}}{16V} + \dfrac{\sqrt{M^2V}\sqrt{28}}{28} - \dfrac{4\sqrt{63}\sqrt{16V}}{16V}$

$= \dfrac{4R\sqrt{7V}}{16V} + \dfrac{2M\sqrt{7V}}{28} - \dfrac{48\sqrt{7V}}{16V}$ $\therefore$ answer $= \underline{\sqrt{7V}\left(\dfrac{R}{4V} + \dfrac{M}{14} - \dfrac{3}{V}\right)}$

PROBLEMS 20-6

2. $\sqrt{16} = \pm 4$

4. $32\sqrt{75} = \pm 160\sqrt{3}$

6. $\sqrt{144} = \pm 12$

8. $\dfrac{\sqrt{7}\sqrt{16}}{16} \times \dfrac{\sqrt{21}\sqrt{3}}{3} = \dfrac{4\sqrt{7}\sqrt{3}\sqrt{7}\sqrt{3}}{48} = \pm\dfrac{84}{48} = \pm\underline{\dfrac{7}{4}}$

10. $\varepsilon^2 - 3$

12. $9 + 6\sqrt{5} + 5 = \underline{14 + 6\sqrt{5}}$

14. $2(5) + 10\sqrt{10} + 3\sqrt{10} + 15(2) = \underline{40 + 13\sqrt{10}}$

16. $\dfrac{\sqrt{2x^2 - 8x + 8} \cdot \sqrt{32x^2 + 128x + 128}}{4x^2 + 16x + 16} = \dfrac{\sqrt{2(x-2)^2} \cdot \sqrt{32(x+2)^2}}{4(x+2)^2}$

$\therefore \dfrac{\sqrt{2}(x-2) \times \sqrt{32}\cancel{(x+2)}}{4(x+2)\cancel{(x+2)}} = \dfrac{\sqrt{64}(x-2)}{4(x+2)} = \underline{\dfrac{2(x-2)}{x+2}} = \text{answer}$

18. $8 - 4\sqrt{3} + 4\sqrt{3} - 2(3) = \underline{2}$

20. Factor and cancel for solution:

answer $= \underline{\sqrt{\alpha} + \sqrt{\beta}}$

PROBLEMS 20-7

2. $\dfrac{5}{5 - \sqrt{3}} \times \dfrac{5 + \sqrt{3}}{5 + \sqrt{3}} = \dfrac{25 + 5\sqrt{3}}{22}$

4. $\dfrac{7}{3\sqrt{5} + 2} \times \dfrac{3\sqrt{5} - 2}{3\sqrt{5} - 2} = \dfrac{21\sqrt{5} - 14}{45 - 4} = \dfrac{7(3\sqrt{5} - 2)}{41}$

6. $\dfrac{(x + \sqrt{y})(x + \sqrt{y})}{(x - \sqrt{y})(x + \sqrt{y})} = \dfrac{x^2 + 2x\sqrt{y} + y}{x^2 - y}$

8. $\dfrac{(3 - \sqrt{5})(2 - \sqrt{5})}{4 - 5} = \dfrac{6 - 5\sqrt{5} + 5}{-1} = \underline{5\sqrt{5} - 11}$

10. $\dfrac{(\sqrt{R} + \sqrt{Z})(\sqrt{R} + \sqrt{Z})}{R - Z} = \dfrac{R + 2\sqrt{RZ} + Z}{R - Z}$

12. $\dfrac{20 + j18}{3 + j5} \times \dfrac{3 - j5}{3 - j5} = \dfrac{60 - j46 - j^2 90}{9 - j^2 25} = \dfrac{(60 + 90) - j46}{9 + 25} = \dfrac{150 - j46}{34}$

PROBLEMS 20-8

2. $+j\sqrt{100} = j10$

4. $j\sqrt{Y^2} = jY$

6. $-j\sqrt{49\omega^2} = -j7\omega$

8. $j\sqrt{\dfrac{Q^4}{\omega^2 L^2}} = j\dfrac{Q^2}{\omega L}$

10. $2\sqrt{-1} \times \sqrt{16} \times \sqrt{3} = j8\sqrt{3}$

12. $-j\dfrac{13}{\alpha}$

14. $-j\lambda\sqrt{\pi}$

PROBLEMS 20-9

2. $28 + j14$
4. $128 - j27$
6. $37 + j6$
8. $57 - j6$
10. $-4 + j0$
12. $64 - j17$
14. $27 - j6$
16. $-5 - j6$

PROBLEMS 20-10

2. $2 + j39$
4. $-31 - j53$
6. $G^2 + BL + jG(L - B)$

8. $\dfrac{6(1 + j2)}{(1 - j2)(1 + j2)} = \dfrac{6 + j12}{5}$

10. $\dfrac{1 - j2 + j^2 1}{1 - j^2 1} = -j1$

12. $0.5 + j0.5$

14. $\dfrac{(\theta + j\phi)^2}{\theta^2 + \phi^2} = \dfrac{\theta^2 - \phi^2 + j2\theta\phi}{\theta^2 + \phi^2}$

16. $\dfrac{9 + j6}{13}$

18. $\dfrac{\left(1 + j\dfrac{\omega}{\omega_0}\right)\left(1 + j\dfrac{\omega}{\omega_0}\right)}{1 + \left(\dfrac{\omega}{\omega_0}\right)^2} = \dfrac{1 - \left(\dfrac{\omega}{\omega_0}\right)^2 + j2\dfrac{\omega}{\omega_0}}{1 + \left(\dfrac{\omega}{\omega_0}\right)^2}$

20. $\dfrac{1 + j\omega T_1 + j\omega T_2 - \omega^2 T_1 T_2}{\mu_0 - \beta} = \dfrac{1 - \omega^2 T_1 T_2}{\mu_0 - \beta} + j\dfrac{\omega\left(T_1 + T_2\right)}{\mu_0 - \beta}$

PROBLEMS 20-11

2. $R = 16$
4. $p = 4$
6. $L + 5 = 36 \therefore L = 31$

8. $\sqrt{\alpha - 6} = 6$
$\therefore \alpha = 42$

10. $\dfrac{7k + 4}{2} = 16 \therefore k = 4$

12. $E^2 = \dfrac{\eta\phi}{\omega^2\theta} \quad \therefore \phi = \dfrac{E^2\omega^2\theta}{\eta}$

14. $\left(\dfrac{i_s}{i_n}\right)^2 = \dfrac{\rho P_s}{e(\Delta f)} \quad \therefore P_s = \left(\dfrac{i_s}{i_n}\right)^2 \cdot \dfrac{e(\Delta f)}{\rho}$

16. $\dfrac{\lambda^2\gamma^2Q^2}{16\pi^2} = \dfrac{KFTS(\Delta f)}{NP_0}$ $\quad \therefore \dfrac{S}{N} = \dfrac{\lambda^2\gamma^2Q^2P_0}{16\pi^2KFT(\Delta f)}$

and $P_0 = \dfrac{S16\pi^2KFT(\Delta f)}{\lambda^2\gamma^2Q^2N}$

18. $\gamma^2 = \dfrac{1 - \mu_x\eta E}{\omega X}$

$\gamma^2\omega X = 1 - \mu_x\eta E$

$\mu_x\eta E = 1 - \gamma^2\omega X$

$\therefore \mu_x = \dfrac{1 - \gamma^2\omega X}{\eta E}$

20. $Z_t = \sqrt{R^2 + R^2\left(\dfrac{f}{f_0}\right)^4}$

$Z_t^2 = R^2 + R^2\left(\dfrac{f}{f_0}\right)^4$

$\dfrac{Z_t^2 - R^2}{R^2} = \left(\dfrac{f}{f_0}\right)^4$

$\sqrt[4]{\dfrac{Z_t^2 - R^2}{R^2}} = \dfrac{f}{f_0}$ $\quad \therefore f_0 = f\sqrt[4]{\dfrac{R^2}{Z_t^2 - R^2}}$

22. $f = \dfrac{1}{2\pi\sqrt{LC}}$

24. $f = \dfrac{1}{2\pi\sqrt{47 \times 15 \times 10^{-18}}} = \underline{6\ \text{MHz}}$

26. <u>No answer provided.</u>

28. $\dfrac{1}{\tau_1\tau_2} - \dfrac{1}{4\tau_2^2} = (786)^2,$ $\quad \dfrac{1}{\tau_1\tau_2} = (786)^2 + \dfrac{1}{(2\tau_2)^2}$

$\therefore \dfrac{1}{\tau_1} = \tau_2[(786)^2 + (78.6)^2] = (786)^2 + (78.6)^2$

$\tau_1 = \dfrac{1}{\tau_2[(786)^2 + (78.6)^2]}$

$\tau_1 = \dfrac{78.6 \times 2}{6.2397396 \times 10^5} = \underline{2.5 \times 10^{-4}}$

PROBLEMS 21-1

2. $\ell = \pm 6$

4. $\theta = \pm 0.5$

6. $\phi = \pm 0.04$

8. $I^2 = \dfrac{144}{49}$ $\quad \therefore I = \pm\dfrac{12}{7}$

10. $x = \pm 0.15$

12. $\dfrac{28}{R^2 - 9} = \dfrac{(R + 3)^2 - (R^2 - 9) + (R - 3)^2}{R^2 - 9}$

$28 = R^2 + 6R + 9 - R^2 + 9 + R^2 - 6R + 9$

$28 = R^2 + 27$

$\therefore R^2 = 28 - 27$

$R^2 = 1$

$\therefore R = \pm 1$

14. $24a^2 - 18a + 18a - 48 = 0$

$24a^2 = 48$

$\therefore \underline{a = \pm\sqrt{2}}$

PROBLEMS 21-2

2. $(v + 1)(v - 7) = 0$

$\therefore \underline{v = -1}$ or $\underline{v = 7}$

4. $(x - 2)(x - 3) = 0$

$\therefore \underline{x = 2}$ or $\underline{x = 3}$

6. $(\psi - 12)(\psi - 5) = 0$

$\therefore \underline{\psi = 12}$ or $\underline{\psi = 5}$

8. $(\varepsilon - 11)(\varepsilon + 5) = 0$

$\therefore \underline{\varepsilon = 11}$ or $\underline{\varepsilon = -5}$

10. $\dfrac{8 + k^2 + 2k}{k} = \dfrac{2 - 3k}{k}$

$k^2 + 5k + 6 = 0$

$(k + 3)(k + 2) = 0$

$\therefore \underline{k = -3}$ or $\underline{k = -2}$

12. $\dfrac{160}{I^2} = \dfrac{26 - I}{I}$

$160 = 26I - I^2$

$I^2 - 26I + 160 = 0$

$(I - 10)(I - 16) = 0$

$\therefore \underline{I = 16}$ or $\underline{I = 10}$

14. $\dfrac{2F - 6}{17 - F} = \dfrac{F - 2 - 2}{F - 2}$

$2F^2 - 10F + 12 = 21F - F^2 - 68$

$3F^2 - 31F + 80 = 0$

$F^2 - 10\frac{1}{3}F + 26\frac{2}{3} = 0 \qquad \therefore (F - 5)\left(F - 5\frac{1}{3}\right) = 0$ and $\underline{F = 5}$ or $\underline{F = 5\frac{1}{3}}$

PROBLEMS 21-3

2. $X^2 - 2X = 48$

$X^2 - 2X + 1 = 48 + 1$

$(X - 1)^2 = 49$

$X - 1 = \pm 7$

$\therefore \underline{X = 8}$ or $\underline{X = -6}$

4. $\Omega^2 + 5\Omega = -6$

$\Omega^2 + 5\Omega + \frac{25}{4} = -6 + \frac{25}{4}$

$\Omega + \frac{5}{2} = \pm\sqrt{\frac{1}{4}}$

$\therefore \underline{\Omega = -3}$ or $\underline{\Omega = -2}$

6. $a^2 + 2a = 63$

$a^2 + 2a + 1 = 64$

$(a + 1)^2 = 64$

$a = -1 \pm \sqrt{64}$

$\therefore \underline{a = 7}$ or $\underline{a = -9}$

8. $e^2 - e = 6$

$e^2 - e + \frac{1}{4} = 6 + \frac{1}{4}$

$e - \frac{1}{2} = \pm\sqrt{\frac{25}{4}}$

$\therefore \underline{e = -2}$ or $\underline{e = 3}$

10. $m^2 + 5m = -6$

$m^2 + 5m + \frac{25}{4} = -6 + \frac{25}{4}$

$4m^2 + 20m + 25 = -24 + 25$

$4m^2 + 20m + 25 = 1$

$(2m + 5)^2 = 1$

$2m + 5 = \pm 1$

$2m = -4$ or -6

$\therefore \underline{m = -2}$ or $\underline{m = -3}$

12. $I^2 - 15I = -26$

$I^2 - 15I + \left(\frac{15}{2}\right)^2 = \left(\frac{15}{2}\right)^2 - 26$

$\left(I - \frac{15}{2}\right)^2 = \frac{225 - 104}{4}$

$I = \frac{15}{2} \pm \sqrt{\frac{121}{4}}$

$I = \frac{15}{2} \pm \frac{11}{2}$

$\therefore \underline{I = 13}$ or $\underline{I = 2}$

14. $\ell^2 + 12\ell + 35 = 0$

$\ell^2 + 12\ell + 36 = 36 - 35$

$(\ell + 6)^2 = 1$

$\therefore \ell = -6 \pm \sqrt{1}$

$\underline{\ell = -7}$ or $\underline{\ell = -5}$

16. $(Z + 2)(Z - 1) = (Z + 1)(-5Z - 14)$

$6Z^2 + 20Z = -12$

$Z^2 + \frac{10}{3}Z = -2$ and $Z^2 + \frac{10}{3}Z + \left(\frac{5}{3}\right)^2 = \left(\frac{5}{3}\right)^2 - 2$

$\therefore \left(Z + \frac{5}{3}\right)^2 = \frac{25 - 18}{9}$ or $Z = -\frac{5}{3} \pm \sqrt{\frac{25 - 18}{9}}$

$= -\frac{5}{3} \pm \sqrt{\frac{7}{9}} \qquad \therefore Z = \frac{-5 \pm \sqrt{7}}{3}$

PROBLEMS 21-4

2. $x = \frac{-2 \pm \sqrt{4 + 60}}{2} = \frac{-2 \pm 8}{2}$ $\qquad \therefore \underline{x = 3}$ or $\underline{x = -5}$

4. $\omega = \frac{6 \pm \sqrt{36 + 28}}{2} = \frac{6 \pm 8}{2}$ $\qquad \therefore \underline{\omega = 7}$ or $\underline{\omega = -1}$

6. $I = \frac{7 \pm \sqrt{49 - 24}}{6} = \frac{7 \pm \sqrt{25}}{6}$ $\qquad \therefore \underline{I = 2}$ or $\underline{I = \frac{1}{3}}$

8. $5R + 10 - 2R^2 + 2R = 0$

$-2R^2 + 7R + 10 = 0$ $\qquad \therefore R = \frac{-7 \pm \sqrt{49 + 80}}{-4} = \underline{\frac{+7 \pm \sqrt{129}}{+4}}$

10. $14I_1^2 + 5I_1 - 1 = 0,\quad I_1 = \frac{-5 \pm \sqrt{25 + 56}}{28}$ $\qquad \therefore I_1 = \frac{1}{7}$ or $I_1 = -\frac{1}{2}$

12. $2(\lambda - 2) = (\lambda + 3)(5 - \lambda)$ $\qquad \therefore \lambda^2 = 19,\ \underline{\lambda = \pm\sqrt{19}}$

14. $-2I^2 - 11I + 21 = 0$

$I = \frac{11 \pm \sqrt{121 + 168}}{-4}$

$I = \frac{11 \pm 17}{-4}$

$\underline{\therefore I = -7 \text{ or } I = \frac{3}{2}}$

16. $-12E^2 + 56E + 43 = 0$

$E = \frac{-56 \pm \sqrt{3136 + 2064}}{-24}$

$E = \frac{-56 \pm \sqrt{5200}}{-24}$

$E = \frac{-56 \pm 20\sqrt{13}}{-24}$

$\underline{E = \frac{-14 \pm 5\sqrt{13}}{-6}}$

PROBLEMS 21-7

2. $a(a + 2) = 168$

$a^2 + 2a - 168 = 0$

$a = \frac{-2 \pm \sqrt{4 + 672}}{2}$

$= \frac{-2 \pm 26}{2}$

$\therefore \underline{a = 12},\ \underline{a + 2 = 14}$

$\underline{\text{(Note: } a \neq -14)}$

4. $a^2 + (a + 1)^2 = (a + 2)^2$

$a^2 + a^2 + 2a + 1 = a^2 + 4a + 4$

$a^2 - 2a - 3 = 0$

$\therefore a = \frac{2 \pm \sqrt{16}}{2},\ a = \frac{2 \pm 4}{2}$

$\underline{\therefore a = 3\ (a \neq -1)}$

If $a^2 + b^2 = c^2$ and $b = a + 1$
and $c = a + 2$,
then $a = 3$, $b = 4$, and $c = 5$

6. $x^2 + x - 182 = 0$

$$x = \frac{-1 \pm \sqrt{1 + 728}}{2}$$

$$= \frac{-1 \pm 27}{2} \qquad \therefore \underline{x = 13} \text{ or } \underline{x = -14}$$

8. (a) $V = \pm \sqrt{\frac{32Fr}{w}}$

$$V = \pm 4\sqrt{\frac{2Fr}{w}}$$

(b) F would increase by a factor of four.

(c) $w = \frac{F32r}{V^2} = \frac{32 \times 1.75 \times 15}{22^2}$

$$\therefore w = \frac{16 \times 1.75 \times 15}{11 \times 22}$$

$$\therefore \underline{w = 1.74}$$

10. $(R_t + 1)\left(\frac{do}{di}\right)^2 = r$

$$\therefore \frac{do}{di} = \sqrt{\frac{r}{R_t + 1}}$$

12. $\frac{1}{C_1} = \frac{R_a R_1}{C_2(R_b - R_2)}$ and $\frac{1}{C_2} = \frac{R_b - R_2}{C_1 R_a R_1}$

$$\omega^2 = \frac{R_a R_1}{R_1 R_2 C_2^2 (R_b - R_2)} \qquad \omega^2 = \frac{R_b - R_2}{C_1^2 R_1^2 R_a R_2}$$

$$C_2^2 = \frac{R_a}{\omega^2 R_2 (R_b - R_2)} \qquad C_1^2 = \frac{R_b - R_2}{\omega^2 R_1^2 R_a R_2}$$

$$\underline{C_2 = \pm \frac{1}{\omega}\sqrt{\frac{R_a}{R_2(R_b - R_2)}}} \qquad \underline{C_1 = \pm \frac{1}{\omega R_1}\sqrt{\frac{R_b - R_2}{R_a R_2}}}$$

14. $\frac{1}{4}t^2 + 3t - d = 0 \qquad \therefore \underline{t = -6 \pm 2\sqrt{9 + d} \text{ s}}$

16. $1500 = 1176t - 4.9t^2 \quad \therefore 4.9t^2 - 1176t + 1500 = 0$

$$t = \frac{1176 \pm \sqrt{1\ 382\ 976 - 294\ 000}}{9.8}$$

(a) t rising $= \dfrac{1176 - 1043.5}{9.8} = \underline{13.5\ \text{s}}$ (rising)

(b) t falling $= \dfrac{1176 + 1043.5}{9.8} = \underline{226.5\ \text{s}}$ (falling)

(c) Max height at time t = T.P $\longrightarrow \left(\dfrac{-b}{2a}\right) = \dfrac{1176}{9.8} = \underline{120\ \text{s}}$

(d) $\therefore$ Max height $= -(4.9)(120)^2 + 1176(120)$

$= -70\ 560 + 141\ 120$

$= \underline{70.56\ \text{km}}$

18. $I^2(R^2 + \omega^2L^2) = V^2 \quad \therefore R^2 = \dfrac{V^2}{I^2} - \omega^2L^2$

$$\therefore R = \pm\sqrt{\left(\frac{V}{I}\right)^2 - \omega^2L^2} = \pm\sqrt{\left(\frac{140}{1.7}\right)^2 - (200\pi)^2(0.125)^2}$$

$R = \pm\sqrt{6782.0069 - 6168.5027} \quad \therefore \underline{R = 24.8\ \Omega}$ (three significant figures)

(Note: R can only be positive; author's calculator shows 24.8 Ω.
Comment: To nine decimal readout shows 24.769 016 31 ± 10%?)

20. $B = \dfrac{X}{R^2 + X^2} \quad \therefore R = \pm\sqrt{\dfrac{X}{B} - X^2}$

$R = \pm\sqrt{\dfrac{100}{0.008} - 10\ 000} = \pm\sqrt{2500} \quad \therefore \underline{R = 50\ \Omega}$

22. No answer provided.

24. $I = \dfrac{15 \pm\sqrt{225 - 100}}{10} \quad \therefore \underline{I = 2.618\ \text{A}}$ or $\underline{I = 382\ \text{mA}}$

When $I = 2.618$ A: $R_3 = \dfrac{5}{2.618} = \underline{1.91\ \Omega}$ and $R_2 = \dfrac{5}{2.618^2} = \underline{0.73\ \Omega}$

When $I = 380$ mA: $R_3 = \dfrac{5}{0.382} = \underline{13.1\ \Omega}$ and $R_2 = \dfrac{5}{(0.382)^2} = \underline{34.3\ \Omega}$

$\therefore R_3 = \underline{1.91\ \Omega \text{ or } 13.1\ \Omega}$ and $R_2 = \underline{0.73\ \Omega \text{ or } 34.3\ \Omega}$

26. $R_{AC} = 25\text{ k}\Omega \qquad \therefore R_{BC} = 25\text{ k}\Omega - R_{AB}$ and $R_{AB} = 25\text{ k}\Omega - R_{BC}$.

$V = 90\text{ V},\ V_{AB} = 30\text{ V}$

$R_P = R_1 // R_{BC} \qquad \therefore R_P = \dfrac{5KR_{BC}}{5K + R_{BC}}$ but $R_{BC} = 25\text{ k}\Omega - R_{AB}$

$$\therefore R_P = \frac{5\text{k}(25\text{k} - R_{AB})}{5\text{k} + (25\text{k} - R_{AB})} = \frac{(125)\text{k}^2 - 5\text{k}R_{AB}}{30\text{k} - R_{AB}}$$

$$R_T = R_{AB} + R_P = R_{AB} + \frac{(125)\text{k}^2 - 5\text{k}R_{AB}}{30\text{k} - R_{AB}}$$

$$R_T = \frac{30\text{k}R_{AB} - R_{AB}^2 + (125)\text{k}^2 - 5\text{k}R_{AB}}{30\text{k} - R_{AB}} \qquad I_T = \frac{V}{R_T} \text{ and } V = 90\text{ V}$$

$$\therefore I_T = \frac{90(30\text{k} - R_{AB})}{-R_{AB}^2 + 25\text{k}R_{AB} + (125)\text{k}^2} \text{ and } I_T = I_{AB}$$

Since $I_T R_{AB} = V_{AB}$, $\quad 30 = \dfrac{(2700\text{k} - 90R_{AB})(R_{AB})}{-R_{AB}^2 + 25\text{k}R_{AB} + (125)\text{k}^2}$

$$30(-R_{AB}^2 + 25\text{k}R_{AB} + (125)\text{k}^2 = 2700\text{k}R_{AB} - 90R_{AB}^2$$

$$60R_{AB}^2 - 1950\text{k}R_{AB} + (3750)\text{k}^2 = 0$$

Divide by 60 and solve for R_{AB}:

$$R_{AB}^2 - 32\,500R_{AB} + 625 \times 10^5 = 0$$

$$\therefore R_{AB} = \frac{32\,500 \pm \sqrt{(32\,500)^2 - 4(625 \times 10^5)}}{2} = \frac{32\,500 \pm \sqrt{806.25 \times 10^6}}{2}$$

Use $-\sqrt{\ }$ only, since $+\sqrt{\ }$ gives $R_{AB} = 30.45\text{ k}\Omega$, R_{AC} only 25 kΩ

$\therefore R_{AB} = 2052.73\ \Omega$

$R_{BC} = 25\text{ k}\Omega - R_{AB} = 22\,947.3\ \Omega$ and $R_P = 4105.5\ \Omega$

$R_T = R_{AB} + R_P = 6158.19\ \Omega, \qquad I_{AB} = \dfrac{90}{R_T} = 14.6\text{ mA}$

$I_{R_1} = \dfrac{60}{5}\text{ mA}$ and $I_T = I_{BC} + I_{R_1}$

$$\therefore I_{BC} = 14.6 - 12\text{ mA} = 2.6\text{ mA}$$

Answer: $\underline{R_{AB} = 2052.73\ \Omega}$, $\underline{I_{BC} = 2.6\text{ mA}}$

28. Added series capacitor = $\underline{C_2 = 600 \text{ pF}}$

Solution:

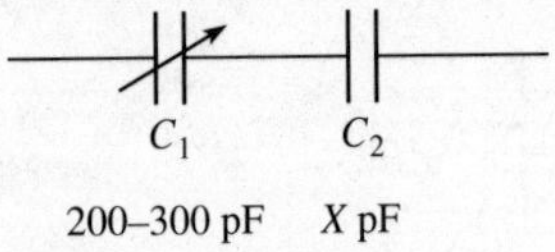

$$\frac{300C_2}{300 + C_2} - \frac{200C_2}{200 + C_2} = 50 \text{ pF}$$

200 - 300 pF

$$\therefore 50 \text{ pF} = \frac{300C_2(200 + C_2) - 200C_2(300 + C_2)}{(300 + C_2)(200 + C_2)}$$

Expanding and cross multiplying:

$$60\ 000C_2 + 300C_2^2 - 60\ 000C_2 - 200C_2^2 = 50 \text{ (denominator)}$$

$$\therefore 100C_2^2 = 50(60\ 000 + 500C_2 + C_2^2)$$

$$100C_2^2 = 3\ 000\ 000 + 25\ 000C_2 + 50C_2^2$$

$$\therefore C_2^2 = 60\ 000 + 500C_2$$

$$C_2^2 - 60\ 000 - 500C_2 = 0$$

Factor: $(C_2 - 600)(C_2 + 100) = 0$ $\quad \therefore \underline{C_2 = 600 \text{ pF}}$ $(C \neq -100 \text{ pF})$

PROBLEMS 22-1

2. $\dfrac{1}{R_T} = \dfrac{1}{2.7k} + \dfrac{1}{1.8k} + \dfrac{1}{4.7k} + 1.2k$ $\quad \therefore \underline{R_P = 2.078 \text{ k}\Omega}$

$$\therefore I = \frac{120}{2.078} \times 10^{-3} = \underline{57.7 \text{ mA}}$$

4. $IR_T = V_G = 10(0.7 + 3 + 5 + 2 + R_4) = 117$

$107 + 10R_4 = 117$ $\quad \therefore \underline{R_4 = 1\ \Omega}$

6. 0 gauge copper = 1.297 Ω/km

Feeder cable = 2 × 400 m $\quad \therefore R_F = 1.0376\ \Omega$

$$V_{\text{FEED}} = 20 \times 1.0376 = \underline{20.752 \text{ V}}$$

$$\therefore V_G = 660 + 20.752 = \underline{681 \text{ V}}$$

8. (a) $I(2.0 + 0.06 + 1.6 + 5 + 0.03 + 3 + 1.5 + 0.15) = 28 + 6$

$$13.34I = 34$$

$$\underline{I = 2.55 \text{ A}}$$

(b) $V_{\text{gen}} = 0.15 \times 2.55 = 0.38 \text{ V}$

$\therefore$ generator terminal voltage = $\underline{27.6 \text{ V}}$

10. $IR_T = V_T$

$\therefore I(5 + 0.06 + 10 + 0.2 + 3 + 0.1 + 7 + 0.05) = 18 - 6 - 32 + 20$

$25.4I = 0 \qquad \therefore \underline{I = \text{zero}}$

PROBLEMS 22-2

2. $\dfrac{1}{1.5} + \dfrac{1}{1.8} + \dfrac{1}{2.2} = \dfrac{1}{R_P} = \dfrac{5.94}{9.96} \qquad \therefore R_P = 0.596\ 39\ \Omega$

$V = IR_T = 10.4 \times 0.596\ 39 = \underline{6.20\ \text{V}}$

4. $\dfrac{1}{R_T} = \dfrac{1}{150} + \dfrac{1}{50} + \dfrac{1}{100} + \dfrac{1}{200} \qquad \therefore R_P = 24\ \Omega$

$V = IR_T = 2 \times 24 = \underline{48\ \text{V}}$

6. $I_L = 20.87$ A, half through each battery; $P = (10.43)^2(0.15) = \underline{16.3\ \text{W}}$

8. (a) $R_T = 5 + 0.6 + 0.8//0.8$

$\therefore R_T = 6\ \Omega \qquad I_T = \dfrac{24 - 12}{6} = 2\ \text{A} \qquad \left(\text{where } I_T = \dfrac{V}{R_T}\right)$

(b) Current flows from $\underline{b \text{ to } a}$ (electron flow)

PROBLEMS 22-3

2. (a) $I_T = \dfrac{G_2}{R_2} + \dfrac{G_1}{R_1} = \dfrac{110}{39} + \dfrac{117}{47} = \underline{5.31\ \text{A}}$

(b) $P_{G_1} = \dfrac{(117)^2}{47} = \underline{291\ \text{W}}$

4.

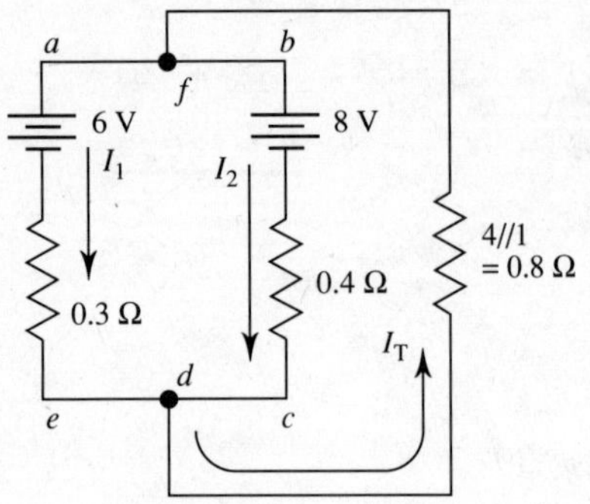

$I_T = I_1 + I_2$

Loop *aedfa*: $1.1I_1 + 0.8I_2 = 6$ Eq. (1)

Loop *bcdfb*: $0.8I_1 + 1.2I_2 = 8$ Eq. (2)

Solve Eq. (1) and Eq. (2) by determinant method for: $I_1 = 1.18$ A,

$I_2 = 5.88$ A.

$\therefore I_T = 7.06$ A.

Voltage across 0.8 Ω = 7.06 × 0.8

= 5.65 V

(a) Power dissipated by 1 Ω = $(5.65)^2$

= <u>31.9 W</u>

(b) Terminal voltage of 6-V battery = <u>5.65 V</u>

6.

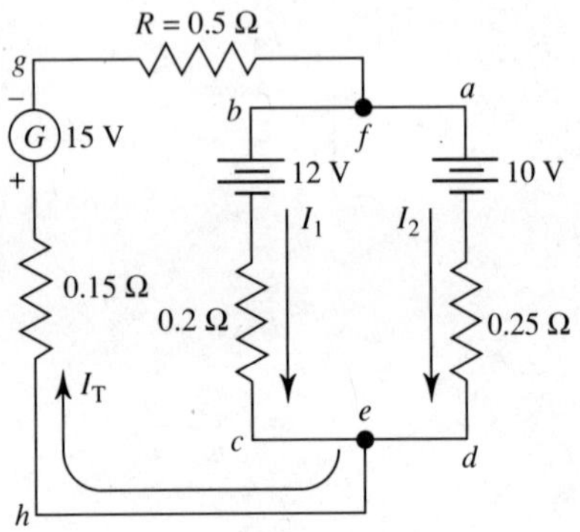

$I_T = I_1 + I_2$

Loop *fadehgf*: $0.65I_1 + 0.9I_2 = 25$ Eq. (1)

Loop *fbcehgf*: $0.85I_1 + 0.65I_2 = 27$ Eq. (2)

Solve I_1 and I_2 by determinants:

$I_1 = 23.504$ A, $I_2 = 10.803$ A

$\therefore I_T = 34.307$ A

Power dissipated by $R = (I_T)^2R$

$P = (34.307)^2(0.5) =$ <u>588 W</u>

10-V battery terminal voltage = $10 - 0.25I_2$

= <u>7.3 V</u>

8.

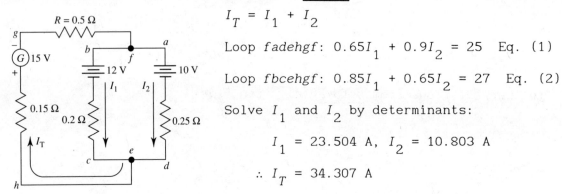

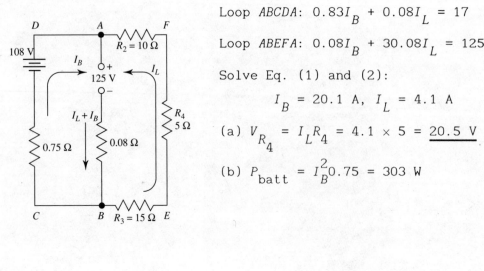

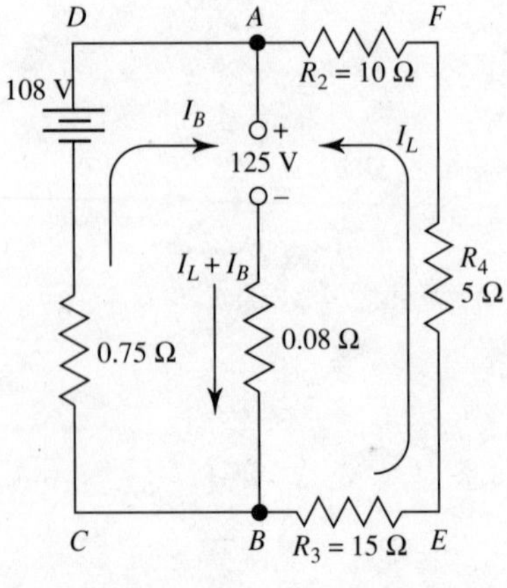

Loop *ABCDA*: $0.83I_B + 0.08I_L = 17$ Eq. (1)

Loop *ABEFA*: $0.08I_B + 30.08I_L = 125$ Eq. (2)

Solve Eq. (1) and (2):

$I_B = 20.1$ A, $I_L = 4.1$ A

(a) $V_{R_4} = I_L R_4 = 4.1 \times 5 =$ <u>20.5 V</u>

(b) $P_{batt} = I_B^2 0.75 = 303$ W

10.

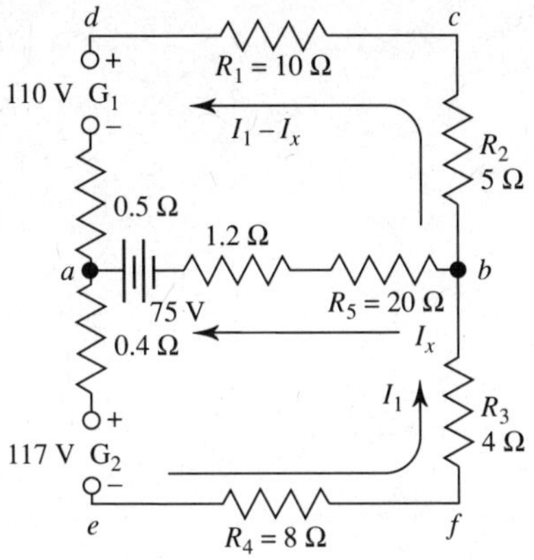

Loop *abcda*: $15.5I_1 - 36.7I_x = 185$ Eq. (1)

Loop *aefba*: $12.4I_1 + 21.2I_x = 42$ Eq. (2)

Solve Eq. (1) and Eq. (2) by determinants:

$\therefore I_1 = 6.97$ A, $I_x = -2.097$ A

*NOTE: Determinant solution results in a negative value for I_x. What does this imply?

(a) $P_{R_5} = (I_x)^2 R_5 = \underline{87.9\ \text{W}}$

(b) $P_{G_2} = (I_1)^2(0.4) = \underline{19.4\ \text{W}}$

12.

(R_b is open cct)

($I_G = I_B + 50$ A)

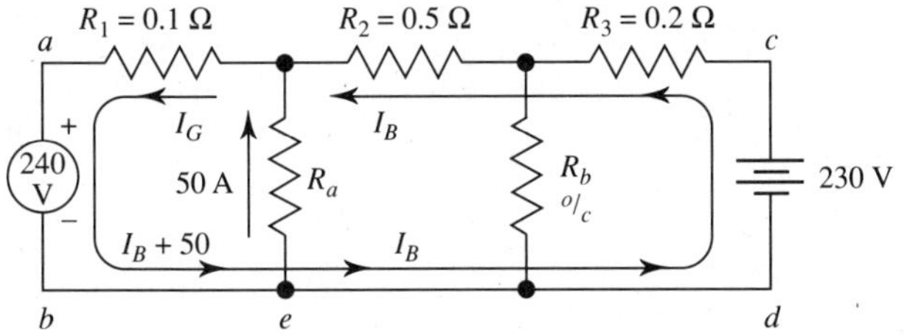

Loop *abdca*: $-240 + 230 + 0.7I_B + 5 + 0.1I_B = 0$

(a) $V_{R_2} = I_B R_2 = (6.25)(0.5) = \underline{3.13\ \text{V}}$ $\qquad 0.8I_B = 5 \qquad \therefore I_B = 6.25$ A

(b) $\underline{I_B = 6.25 \text{ A from } d \text{ to } c}$ (as shown/assumed in diagram)

14.

$R_1 = 0.1\ \Omega$, $R_2 = 0.5\ \Omega$

$R_3 = 0.2\ \Omega$, R_b NOT O/C

$I_G = 150 + I_2$ A

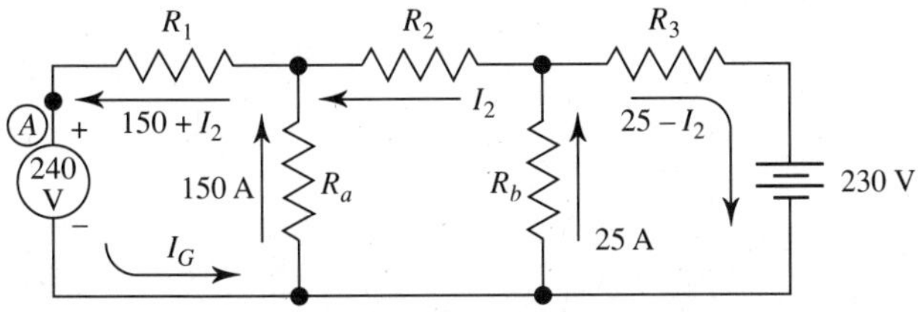

Starting @ point A, in the direction of current (electron movement):

$-240 + 230 - 0.2(25 - I_2) + 0.5I_2 + 0.1(150 + I_2) = 0$

Simplifying: $0.8I_2 = 0 \qquad \therefore \underline{I_2 = 0 \text{ and } P_{R_2} \text{ must be zero}}$

PROBLEMS 22-4

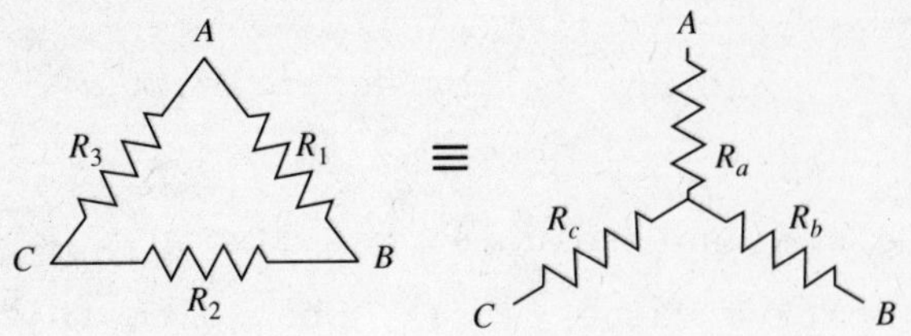

DIAGRAMS FOR PROBLEMS 2, 4, 6, AND 8 (PARTIAL)

2. $R_a = \dfrac{R_3 R_1}{\Sigma\Delta R} = \dfrac{330 \times 150}{700} = \underline{70.7\ \Omega}$, $R_b = \dfrac{220 \times 150}{700} = \underline{47.1\ \Omega}$

$R_C = \dfrac{330 \times 220}{700} = \underline{104\ \Omega}$

4. $\Sigma R_Y = R_a R_b + R_a R_c + R_b R_c = 1325\ \Omega$

$R_1 = \dfrac{\Sigma R_Y}{R_C} = \underline{28.2\ \Omega}$, $R_2 = \dfrac{\Sigma R_Y}{R_a} = \underline{133\ \Omega}$, $R_3 = \dfrac{\Sigma R_Y}{R_b} = \underline{88\ \Omega}$

6. $\Sigma R_Y = 3(1.5)^2 \times 10^3 = 6570\ \text{k}\Omega$, and $R_1 = R_2 = R_3 = \underline{4.5\ \text{k}\Omega}$

8. From Prob. 7: $R_T = 22.57\ \Omega$ and $I_T = 443$ mA (approximately)

$$I_{R_5} = \frac{V_S - V_a}{R_S + R_C} = \frac{10 - (443 \times 10^{-3} \times 10.667)}{15 + 26.667} = \underline{126.6\ \text{mA}}$$

10.

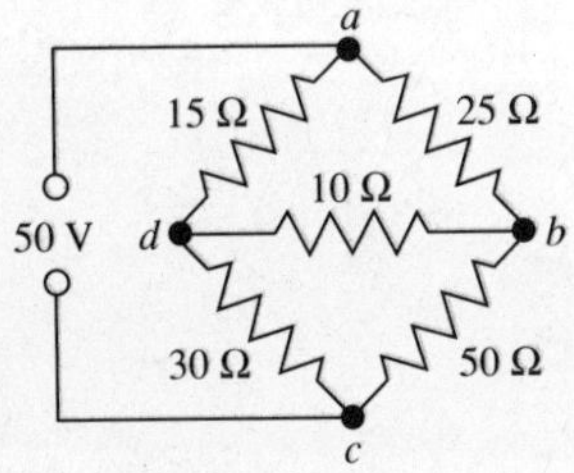

Solve Δ *bcd*:

$R_c = \dfrac{1500}{90} = 16.67\ \Omega$

$R_b = \dfrac{500}{90} = 5.56\ \Omega$

$R_d = \dfrac{300}{90} = 3.33\ \Omega$

$R_T = 16.67 + 18.33//30.56 = 28.13\ \Omega$

$I = \dfrac{V}{R_T} = \dfrac{50}{28.13} = \underline{1.78\ \text{A}}$ (Prove this answer by Kirchhoff's equations.)

12.

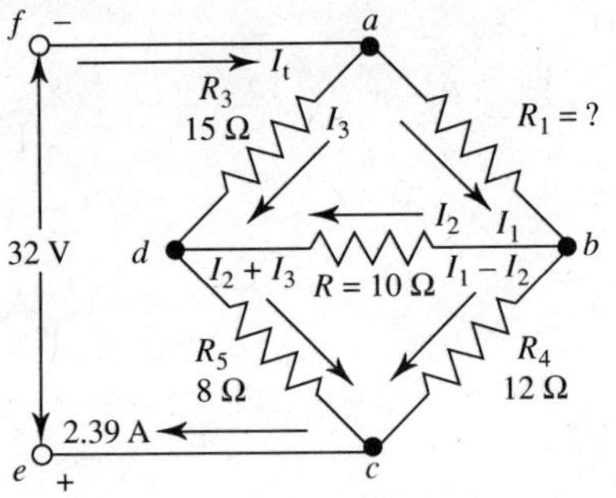

Loop *fadcef*: $32 = 15I_3 + 8I_2 + 8I_3$ Eq. (1)

Loop *fabcef*: $32 = I_1R_1 + 12I_1 - 12I_2$ Eq. (2)

Loop *bdcb*:

$$0 = 10I_2 + 8I_2 + 8I_3 - 12I_1 + 12I_2 \quad \text{Eq. (3)}$$

$$I_t = I_1 + I_3 \text{ and } I_t = 2.39 \text{ A}$$

Substitute for I_3 in Eq. (1):

$$32 = 15(2.39 - I_1) + 8(2.39 - I_1 + I_2)$$

Eq. (2): $32 = I_1R_1 + 12I_1 - 12I_2$

∴ Eq. (4): $32 = I_1R_1 + 10I_2 + 8(2.39 - I_1 + I_2)$

Eq. (5): $-22.97 = -23I_1 + 8I_2$

Eq. (6): $32 = I_1R_1 + 12I_1 - 12I_2$

Eq. (7): $12.88 = I_1R_1 - 8I_1 + 8I_2$

(6) - (7) = $19.12 = 20I_1 - 20I_2$ = Eq. (8) ∴ since $I_3 = I_t - I_1$,

Solve Eq. (5) and Eq. (8) by determinants:

$I_1 = 1.021\,47$ A

and $I_2 = 0.065\,47$ A

$I_3 = I_T - I_1$

$I_1 = 1.021\,47$ A and $I_T = 2.39$ A (given)

∴ $I_3 = 1.368\,53$ A

$I_3R_3 = I_1R_1 + I_2R_2$

∴ $1.368\,53 \times 15 = 1.021\,47R_1 + 0.065\,47 \times 10$

$$\therefore R_1 = \frac{20.527\,95 - 0.6547}{1.021\,47} = \underline{20\ \Omega}\ (19.5)$$

14. Solve delta formed by R_2, R_3, R_4 and redraw circuit:

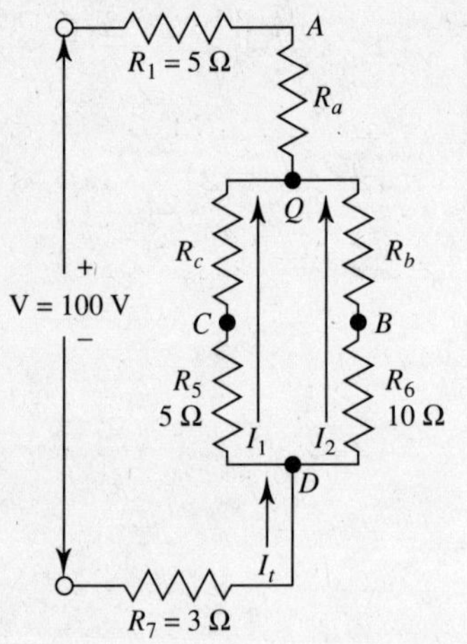

$R_a = 5\ \Omega$, $R_b = 8.33\ \Omega$, $R_c = 6.25\ \Omega$

$$R_T = R_1 + R_a + R_7 + [(R_c + R_5)//(R_b + R_6)]$$

$$= 13 + 11.25//18.33 \qquad \therefore R_T = 20\ \Omega$$

$$I_T = \frac{V}{R_T} = \frac{100}{20} = 5\text{ A}$$

$$V_{DQ} = V - 13I_t = 34.91\text{ V}$$

$$\therefore I_1 = \frac{V_{DQ}}{R_c + R_5} = \frac{34.91}{11.25} = 3.1\text{ A}$$

$$\therefore I_2 = 1.9\text{ A}$$

$$V_{CD} = 5I_1 = 15.5\text{ V}$$

$$V_{BD} = 10I_2 = 19.0\text{ V}$$

$$V_{R_4} = V_{BC} = V_{BD} - V_{CD} = 3.5\text{ V}$$

$$\therefore I_{R_4} = \frac{V_{R_4}}{R_4} = \frac{3.5}{25} = \underline{140\text{ mA}}$$

16.

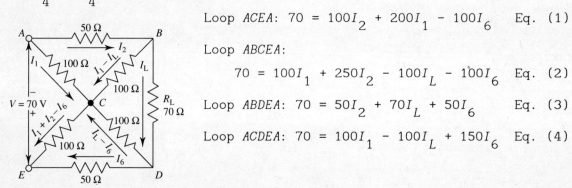

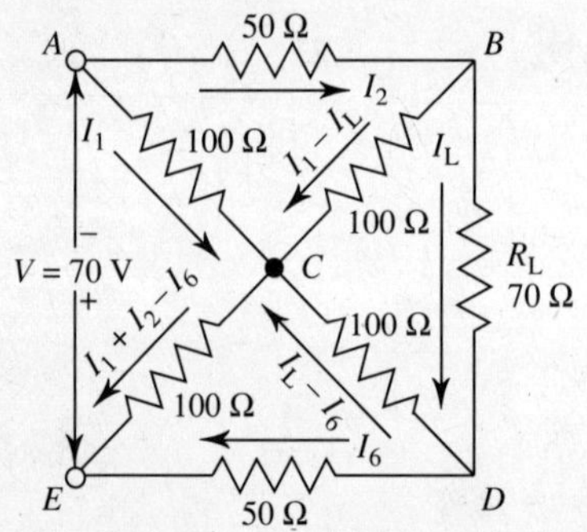

Loop *ACEA*: $70 = 100I_2 + 200I_1 - 100I_6$ Eq. (1)

Loop *ABCEA*:

$70 = 100I_1 + 250I_2 - 100I_L - 100I_6$ Eq. (2)

Loop *ABDEA*: $70 = 50I_2 + 70I_L + 50I_6$ Eq. (3)

Loop *ACDEA*: $70 = 100I_1 - 100I_L + 150I_6$ Eq. (4)

Divide equations by 10 and set up cofactors

Denominator: I_1 I_2 I_L I_6

$$\begin{vmatrix} 20 & 10 & 0 & -10 \\ 10 & 25 & -10 & -10 \\ 0 & 5 & 7 & 5 \\ 10 & 0 & -10 & 15 \end{vmatrix} \begin{matrix} \text{Eq. (1)} \\ \text{Eq. (2)} \\ \text{Eq. (3)} \\ \text{Eq. (4)} \end{matrix}$$

Solution of denominator = 102 500

Set up numerator I_L: $\begin{vmatrix} 20 & 10 & 7 & -10 \\ 10 & 25 & 7 & -10 \\ 0 & 5 & 7 & 5 \\ 10 & 0 & 7 & 15 \end{vmatrix} = 35\ 000$

$$\therefore I_L = \frac{35\ 000}{102\ 500} = \underline{341\text{ mA}}$$

Prove the answer in Prob. 16 using the Δ ⟶ Y method!

PROBLEMS 22-5

2. (a) $V_{TH} = 7.2$ V (b) $I_N = 120$ mA and $R_{TH} = R_N = \dfrac{7.2}{120 \times 10^{-3}} = 60\ \Omega$

Note: Prob. 2 may be described as:

(a) a constant 7.2 V source in series with 60.0 Ω

(b) a constant 120 mA source in parallel with 60.0 Ω

4.

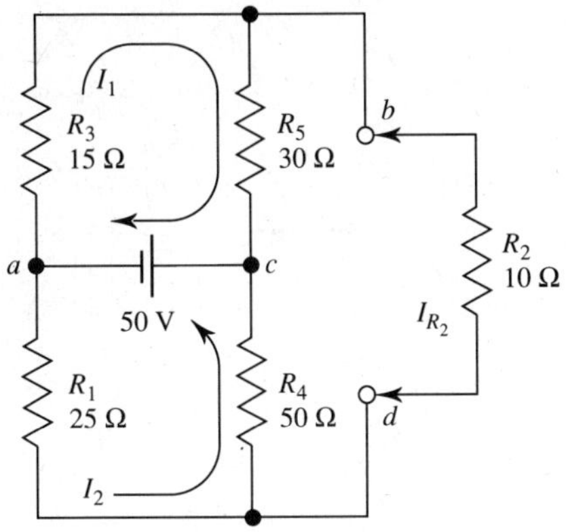

$$I_1 = \frac{50}{45} = 1.11\text{ A} = I_{R_3}$$

$$I_2 = \frac{50}{75} = 0.667\text{ A}$$

$$V_{R_4} = I_2R_4 = 33.3\text{ V}$$

$$V_{R_5} = I_1R_5 = 33.3\text{ V}$$

$$V_{TH} = V_{R_4} - V_{R_5} = 0 \qquad \therefore I_{R_2} = 0$$

$$R_{TH} = \frac{R_5R_3}{R_5 + R_3} = \frac{R_4R_1}{R_4 + R_1} = 10 + 16.7 = 26.7\ \Omega$$

$V_{TH} = 0$, $R_{TH} = 26.7\ \Omega$, $I_{R_3} = 1.11$ A, $I_{R_2} = 0$

6.

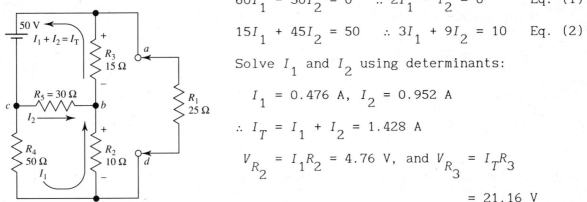

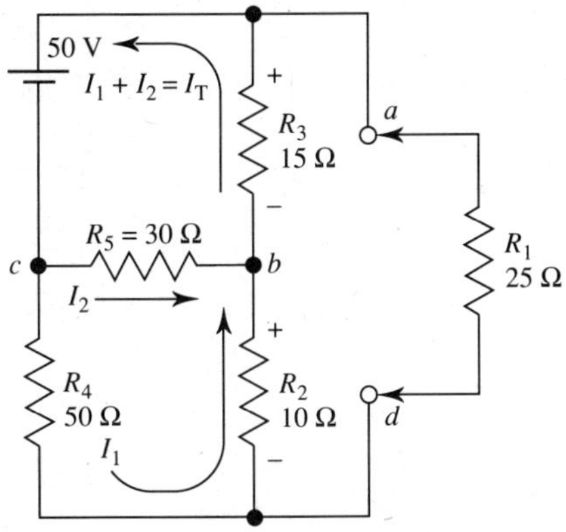

$60I_1 - 30I_2 = 0 \quad \therefore 2I_1 - I_2 = 0$ Eq. (1)

$15I_1 + 45I_2 = 50 \quad \therefore 3I_1 + 9I_2 = 10$ Eq. (2)

Solve I_1 and I_2 using determinants:

$I_1 = 0.476$ A, $I_2 = 0.952$ A

$\therefore I_T = I_1 + I_2 = 1.428$ A

$V_{R_2} = I_1R_2 = 4.76$ V, and $V_{R_3} = I_TR_3$

$= 21.16$ V

$$V_{TH} = V_{R_2} + V_{R_3} = 26.16\text{ V}$$

$$R_{TH} = R_4//[(R_5//R_3) + R_2] = 14.3\ \Omega$$

Thevenin equivalent = 26.16 V in series with 14.3 Ω

$$\therefore I_{R_1} = \frac{V_{TH}}{R_{TH} + R_1} = \frac{26.16}{39.3} = 665\text{ mA}$$

PROBLEMS 23-1

2. (a) 115° (b) 20° (c) -55° (d) -90° (e) -160° (f) 310°

6. (a) $1h = 360^\circ$ $\therefore \frac{20}{60} \times 360 = \underline{120^\circ}$ (b) $\frac{40}{60} \times 360 = \underline{240^\circ}$

8. (a) Second hand = $\underline{360^\circ/\text{min}}$

(b) Minute hand = $\underline{6^\circ/\text{min}}$

(c) Hour hand = $\underline{0.5^\circ/\text{min}}$

10. $C = 300\pi$ mm (or 0.3π m)

1800 rev/min = 30 rev/s

$1s = 30 \times 300\pi = \underline{9\pi \text{ m/s}}$

PROBLEMS 23-2

2. (a) $1^r \times \frac{180^\circ}{\pi^r} = \underline{57.3^\circ}$ (b) $0.5 \times \frac{180^\circ}{\pi} = \underline{28.6^\circ}$

(c) $\frac{1}{\pi} \times \frac{180^\circ}{\pi} = \underline{18.2^\circ}$ (d) $\frac{\pi}{5} \times \frac{180^\circ}{\pi} = \underline{36^\circ}$

(e) $\frac{2\pi}{3} \times \frac{180^\circ}{\pi} = \underline{120^\circ}$ (f) $0.785\ 40 \times \frac{180^\circ}{\pi} = \underline{45^\circ}$

4. Hr hand rotates $2\pi^r$ in 720 min $\therefore$ 40 min $= \frac{40}{720} \times 2\pi = \frac{\pi^r}{9}$

6. (a) Second hand: $2\pi^r/60 \text{ s} = \frac{\pi}{30}$ r/s

(b) Minute hand: $2\pi^r/3600 \text{ s} = \frac{\pi}{1800}$ r/s

(c) Hour hand: $2\pi^r/12 \times 3600 \text{ s} = \frac{\pi}{21\ 600}$ r/s

8. 6 rev/min $= \frac{1}{10}$ rev/s

$= \frac{1}{10} \times 2\pi^r = \frac{\pi}{5}$ r/s

10. $2\pi^r$ in $24 \times 60 \text{ min} = \frac{2\pi}{1440} = \frac{\pi}{720}$ r/min

PROBLEMS 23-3

2. (a) 27°, (b) 45°, (c) 54°, (d) 108°, (e) 135°, (f) 270°

4. (a) 50^g, (b) 33.3^g, (c) 80^g, (d) 300^g, (e) 166.7^g, (f) 100^g

PROBLEMS 23-4

2.

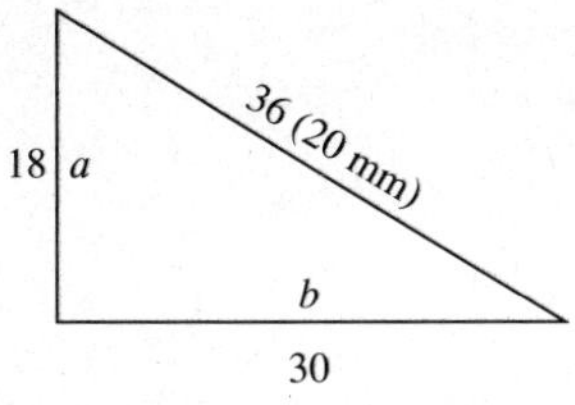

$$\frac{a}{18} = \frac{20}{36} \qquad \therefore\ a = \frac{360}{36} = 10 \text{ mm}$$

$$\frac{b}{30} = \frac{20}{36} \qquad \therefore\ b = \frac{30 \times 20}{36} = 16.67 \text{ mm}$$

4. $A = 36.9^\circ$, $B = 53.1^\circ$, $C = 90^\circ$

6. $a = 8.89$, $B = 62.4^\circ$, $C = 37.6^\circ$

8. $b = 4.95$, $A = 45^\circ$, $C = 90^\circ$

10. $b = 35$, $B = 60^\circ$, $C = 60^\circ$

PROBLEMS 23-5

2. $c = 60$, $B = 36.9^\circ$ (3, 4, 5, triangle)

4. 1 *kn* = 1 nautical mile/h

   In 2h *I* goes 1300 nautical miles North.

   In 2h *H* goes 2200 nautical miles East.

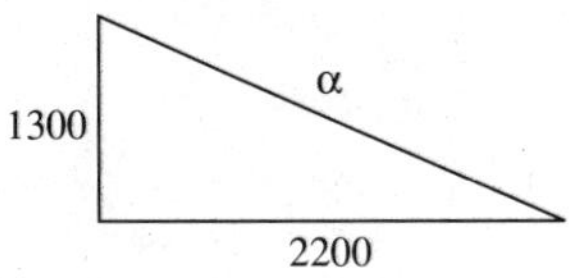

$$\alpha = \sqrt{(1300)^2 + (2200)^2}$$

$$\therefore\ \alpha = 2555.4 \text{ nautical miles}$$

6.

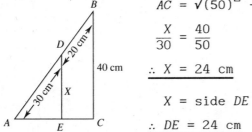

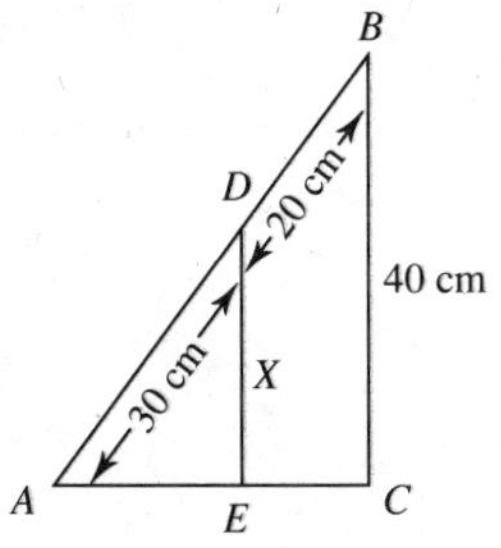

$$AC = \sqrt{(50)^2 - (40)^2} = 30 \text{ cm}$$

$$\frac{X}{30} = \frac{40}{50}$$

$$\therefore\ X = 24 \text{ cm}$$

X = side DE

$$\therefore\ DE = 24 \text{ cm}$$

8.

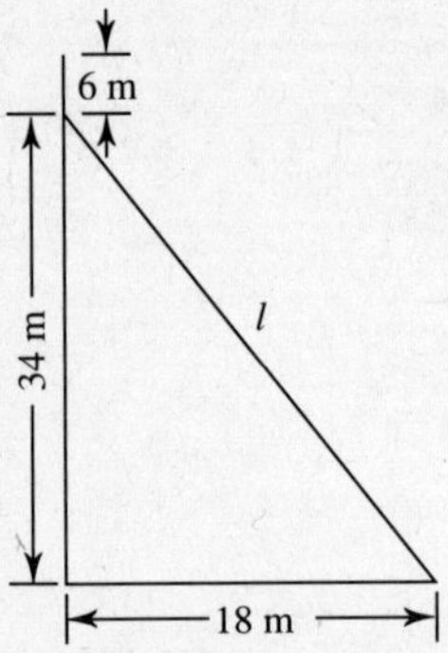

$\ell = \sqrt{(34)^2 + (18)^2}$

$= \sqrt{1480} = \ell = 38.47$

$\therefore \ell = 38.5$ m

10.

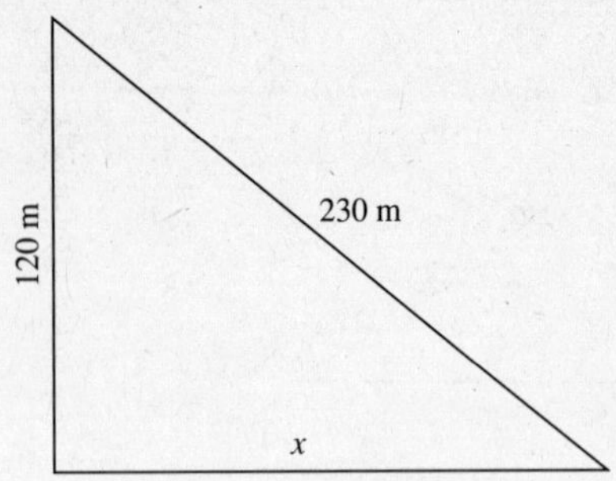

$(230)^2 = x^2 + (120)^2$

$\therefore x = \sqrt{(230)^2 - (120)^2}$

$x = 196.2$ m

PROBLEMS 24-1

2. (a) $\sin\alpha = \frac{OR}{PR}$ (b) $\sin\beta = \frac{OP}{PR}$ (c) $\cot\beta = \frac{OR}{OP}$ (d) $\sec\alpha = \frac{PR}{OP}$

(e) $\tan\alpha = \cot\beta = \frac{OR}{OP}$

4. $\sin\alpha = \frac{5}{13}$ $\cos\alpha = \frac{12}{13}$ $\tan\alpha = \frac{5}{12}$

$\csc\alpha = \frac{13}{5}$ $\sec\alpha = \frac{13}{12}$ $\cot\alpha = \frac{12}{5}$

$\sin\beta = \frac{12}{13}$ $\cos\beta = \frac{5}{13}$ $\tan\beta = \frac{12}{5}$

$\csc\beta = \frac{13}{12}$ $\sec\beta = \frac{13}{5}$ $\cot\beta = \frac{5}{12}$

6. $\cos\theta = \frac{R}{Z} = \frac{0.5Z}{Z} = \underline{0.5}$

$\sin\theta = \frac{X}{Z} = \underline{0.866}$

$\tan\theta = \frac{\sin\theta}{\cos\theta} = \underline{1.73}$

8. (a) $\csc\theta = \frac{3}{2}$ (b) $\cos\alpha = \frac{1}{2}$

(c) $\tan\beta = \frac{8}{7}$ (d) $\sec\phi = \frac{16}{5}$

(e) $\cot\phi = \frac{1}{12}$ (f) $\sin\alpha = \frac{1}{4}$

10. $\cos\alpha = \frac{4}{5}$ $\sin\alpha = \frac{3}{5}$ $\tan\alpha = \frac{3}{4}$

$\sec\alpha = \frac{5}{4}$ $\csc\alpha = \frac{5}{3}$ $\cot\alpha = \frac{4}{3}$

12. (a) $\tan\theta$ (b) $\cot\theta$ (c) $\sec\theta$ (d) $\csc\theta$

PROBLEMS 24-2

2. I or IV **4.** II or IV **6.** II **8.** II **10.** I **12.** II or III

14. $\tan\theta = \frac{5}{12}$ $\therefore \sin\theta = \frac{5}{13}$ and $\cos\theta = \frac{12}{13}$; $\frac{\sin\theta - \csc\theta}{\cot\theta - \sec\theta} = \frac{\frac{5}{13} - \frac{13}{5}}{\frac{12}{5} - \frac{13}{12}}$

$$= -\frac{1728}{1027}$$

Answer = $\underline{-1.68}$

	Angle	Sine	Cosine	Tangent
16.	210°	$-$	$-$	$+$
18.	350°	$-$	$+$	$-$
20.	$\frac{\pi}{3}^r$	$+$	$+$	$+$
22.	-72°	$-$	$+$	$-$

	Sine	Cosine	Tangent	Secant	Cosecant	Cotangent
26.	$\frac{4}{5}$	$\frac{3}{5}$	$\frac{4}{3}$	$\frac{5}{3}$	$\frac{5}{4}$	$\frac{3}{4}$
28.	$\frac{4}{5}$	$-\frac{3}{5}$	$-\frac{4}{3}$	$-\frac{5}{3}$	$\frac{5}{4}$	$-\frac{3}{4}$
30.	$\frac{\sqrt{2}}{2}$	$\frac{\sqrt{2}}{2}$	$+1$	$\sqrt{2}$	$\sqrt{2}$	$+1$
32.	$\frac{3}{5}$	$-\frac{4}{5}$	$-\frac{3}{4}$	$-\frac{5}{4}$	$\frac{5}{3}$	$-\frac{4}{3}$
34.	$\frac{\sqrt{2}}{2}$	$\frac{\sqrt{2}}{2}$	$+1$	$\sqrt{2}$	$\sqrt{2}$	$+1$

PROBLEMS 24-3

2. 0 **4.** ∞ **6.** (a) 1, (b) 1, (c) -1, (d) -1

PROBLEMS 25-1 *NOTE: All sine, cosine, and tangent values have been read/taken to five decimal places from a fully scientific calculator.

		Sine	Cosine	Tangent
2.	(a)	0.207 91	0.978 15	0.212 56
	(b)	0.999 74	0.022 69	44.066 11
	(c)	0.940 88	0.338 74	2.777 61
	(d)	0.013 96	0.999 90	0.013 96
	(e)	0.343 66	0.939 09	0.365 95
4.	(a)	0.128 62	0.991 69	0.129 70
	(b)	0.210 98	0.977 49	0.215 84
	(c)	0.539 51	0.841 98	0.640 76
	(d)	0.663 27	0.748 38	0.886 28
	(e)	0.055 30	0.998 47	0.055 38

PROBLEMS 25-2

2. (a) 15°
(b) 79°
(c) 2.9°
(d) 29.9°
(e) 69.93°

4. (a) 53.6°
(b) 60.9°
(c) 15.9°
(d) 87.2°
(e) 31.2°

PROBLEMS 25-3

		Sine	Cosine	Tangent
2.	(a)	-0.052 34	-0.998 63	0.052 41
	(b)	-0.819 15	-0.573 58	1.428 15
	(c)	-0.612 91	-0.790 16	0.775 68
	(d)	-0.015 71	-0.999 88	0.015 71
	(e)	-0.999 45	-0.033 16	30.144 62
4.	(a)	0.981 63	-0.190 81	-5.144 55
	(b)	0.500 00	-0.866 03	-0.577 35
	(c)	0.861 63	-0.507 54	-1.697 66
	(d)	0.289 03	-0.957 32	-0.301 92
	(e)	0.012 22	-0.999 93	-0.012 22

6. (a) $\theta = 14.3^\circ$ (b) $\theta = 35.6^\circ$
(c) $\theta = 117.1^\circ$ (d) $\theta = -74.1^\circ$ (e) $\theta = -22.2^\circ$

8. $F = \dfrac{Ed^2}{\cos\theta}$ ℓm, $d = \sqrt{\dfrac{F\cos\theta}{E}}$ m, $\theta = \arccos \dfrac{Ed^2}{F}$

10. $E = \dfrac{F\cos\theta}{d^2}$ ℓx $= \dfrac{1700 \cos 60^\circ}{9} = \dfrac{850}{9} = \underline{94.4 \text{ ℓx}}$

12. $F = \dfrac{E_h h^2}{\cos^3\theta}$ ℓm and $\theta = \arccos \sqrt[3]{\dfrac{E_h h^2}{F}}$, $h = \sqrt{\dfrac{\cos^3\theta F}{E_h}}$ m

14. $F = \dfrac{(330)(4.5)^2}{\cos^3 50^\circ} = \underline{2.52 \times 10^4 \text{ ℓm}}$

16. $h = 2$ m, $\theta = 0^\circ$, $\therefore \cos\theta = 1$.

From Prob. 15 E_h is 70% downward, $\therefore E_h = \dfrac{1250}{0.7}$

From Prob. 12:

$$F = \frac{E_h h^2}{\text{Cos}^3\theta} \text{ ℓm} = \frac{1250}{0.7} \times 4 = \underline{7.14 \times 10^3 \text{ ℓm}}$$

PROBLEMS 26-1 Note: Diagram for Problems 26-1 through 26-4 including sample solutions

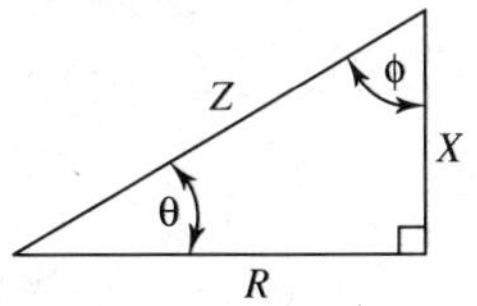

(no scale implied)

2. $\cos\phi = \dfrac{X}{Z}$ $\therefore Z = \dfrac{7.25}{\cos 68.5} = \underline{19.8}$

$\theta = 90 - \phi = \underline{21.5^\circ}$

$R = Z \sin\phi = 19.8 \times 0.9304 = \underline{18.4}$

4.	$Z = 9.21$	$X = 6.63$	$\theta = 46^\circ$
6.	$Z = 1596$	$R = 453$	$\phi = 16.5^\circ$
8.	$Z = 501$	$X = 449$	$\theta = 63.7^\circ$
10.	$Z = 1.02$	$X = 0.996$	$\phi = 13^\circ$
12.	$Z = 0.833$	$R = 0.828$	$\phi = 83.6^\circ$
14.	$Z = 1.08$	$X = 0.850$	$\phi = 38.1^\circ$
16.	$Z = 1.0$	$X = 0.5\sqrt{2}$	$\phi = 45^\circ$

PROBLEMS 26-2 Use diagram above and ($X = Z \sin\theta$), ($R = Z \sin\phi$ or $Z \cos\theta$)

2.	$R = 516.9$	$X = 351.3$	$\phi = 55.8^\circ$
4.	$R = 118.2$	$X = 43.7$	$\theta = 20.3^\circ$
6.	$R = 24.96$	$X = 31.3$	$\phi = 38.6^\circ$
8.	$R = 600$	$X = 109$	$\theta = 10.3^\circ$
10.	$R = 0.327$	$X = 0.0989$	$\theta = 16.8^\circ$

PROBLEMS 26-3 Use diagram on page 87

and ($X = Z \sin \theta$), ($R = Z \sin \phi$ or $Z \cos \theta$)

2. $\theta = 62.1^\circ$ $\phi = 27.9^\circ$ $X = 1627$

4. $\theta = 9.2^\circ$ $\phi = 80.8^\circ$ $X = 496$

6. $\theta = 8.1^\circ$ $\phi = 81.9^\circ$ $R = 403$

8. $\theta = 73.3^\circ$ $\phi = 16.7^\circ$ $X = 38$

10. $\theta = 17^\circ$ $\phi = 73^\circ$ $X = 0.100$

PROBLEMS 26-4

2. $\theta = 49.5^\circ$ $\phi = 40.5^\circ$ $Z = 21.8$

4. $\theta = 69^\circ$ $\phi = 21.0^\circ$ $Z = 237.8$

6. $\theta = 78.5^\circ$ $\phi = 11.5^\circ$ $Z = 51.6$

8. $\theta = 30^\circ$ $\phi = 60^\circ$ $Z = 1$

10. $\theta = 26.2^\circ$ $\phi = 63.8^\circ$ $Z = 4.7$

PROBLEMS 26-5

2. $\tan A = \dfrac{3.68}{1.36} = 2.7059$ $\therefore A = 69.7^\circ$

4. $ht = 174 \tan 41.7^\circ = \underline{155 \text{ m}}$ **6.** $ht = 85 \tan 31^\circ = \underline{51.1 \text{ m}}$

8.

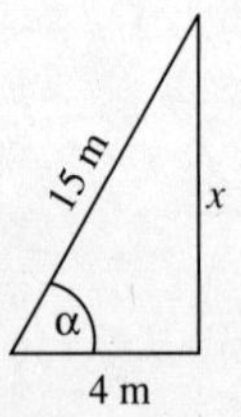

(a) $\cos \alpha = \dfrac{4}{15} = 0.266\,67$ $\therefore \alpha = 74.5^\circ$

(b) $x = 15 \sin \alpha$ m

$= 15 \times 0.963\,79$

$\therefore x = \underline{14.5 \text{ m}}$

10.

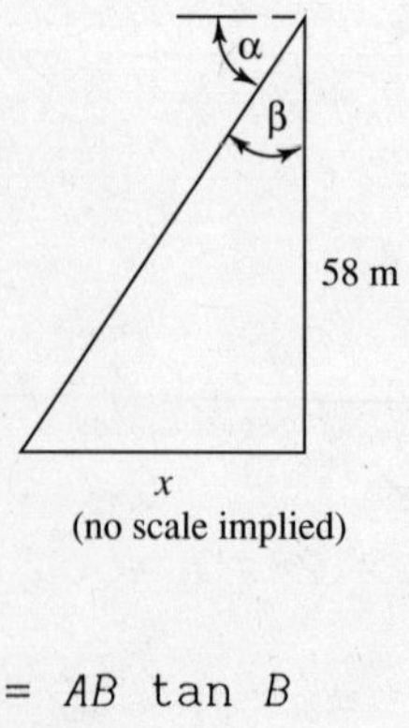

$\alpha + \beta = 90^\circ$

$\alpha = 28.6^\circ$ $\therefore \beta = 90 - 28.6^\circ$ and $\beta = 61.4^\circ$

$\dfrac{x}{58} = \tan \beta$

$\therefore x = 58(\tan 61.4^\circ)$

$= 58(1.834\,13)$ and $x = \underline{106.4 \text{ m}}$

12. $AC = AB \tan B$

$= 240(\tan 59.1^\circ)$

$= 240(1.670\,88)$ $\therefore AC = 401$ m

As an exercise determine distance BC.

PROBLEMS 26-6

2. $a = 77$ mm, $b = 96.4$ mm, $c = 72$ mm

$\theta = 52^\circ$, $\phi = 47.5^\circ$, A = area

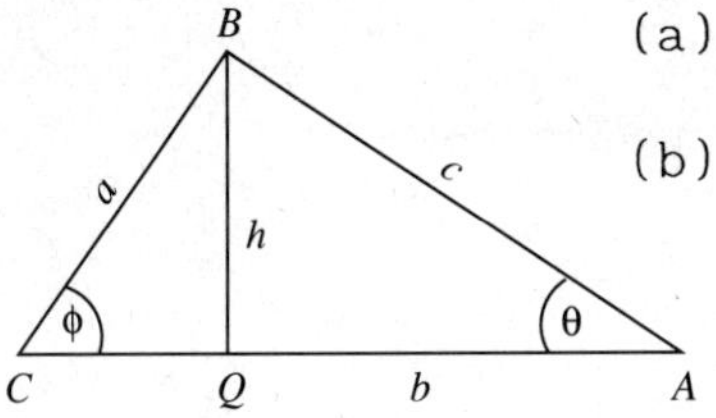

(a) $\angle B = 180^\circ - (\theta + \phi) = \underline{80.5^\circ}$

(b) $A = \frac{1}{2}ab \sin \phi$

$= \frac{1}{2}(77)(96.4)(\sin 47.5^\circ)$

$\underline{\therefore A = 2.74 \times 10^3 \text{ mm}^2}$

(c) $a \sin \phi = h = c \sin \theta = \underline{56.8 \text{ mm}}$

(d) $\frac{1}{2} CQh + \frac{1}{2}AQh$ = area $\qquad CQ = a \cos \phi$, $AQ = c \cos \theta$

$\therefore \frac{1}{2}ah \cos \phi + \frac{1}{2}ch \cos \theta$ = area

$\frac{77 \times 56.8 \cos \phi + 72 \times 56.8 \cos \theta}{2} = \underline{\text{area} = 2.74 \times 10^3 \text{ mm}^2}$

PROBLEMS 27-1

2. $\sec^2 \phi = \tan^2 \phi + 1 \qquad$ If $\tan = \frac{\sin}{\cos}$, then:

$\frac{1}{\cos^2 \phi} = \frac{\sin^2 \phi}{\cos^2 \phi} + 1 \qquad \therefore 1 = \sin^2 \phi + \cos^2 \phi$

4. $(\sin^2 \alpha - \cos^2 \alpha)(\sin^2 \alpha + \cos^2 \alpha) = \sin^2 \alpha - \cos^2 \alpha$

$\sin^2 \alpha + \cos^2 \alpha = \frac{\sin^2 \alpha - \cos^2 \alpha}{\sin^2 \alpha - \cos^2 \alpha}$

$\underline{\therefore \sin^2 \alpha + \cos^2 \alpha = 1}$

6. $\cos^2 \phi = 1 - \sin^2 \phi$

$\underline{\therefore 1 = \sin^2 \phi + \cos^2 \phi}$

The remaining problems are left as exercises, since there are many varied (and valid) methods of solving the identities within this chapter.

PROBLEMS 27-2

Sample solutions:

2. $\alpha = 180^\circ - \beta - \gamma = 180^\circ - 161^\circ = \underline{19^\circ}$, $\underline{c = 97.75}$, $\underline{b = 95.77}$

Thus: $\dfrac{a}{\sin\alpha} = \dfrac{c}{\sin\gamma}$ $\quad\therefore c = \dfrac{a\sin\gamma}{\sin\alpha} = \dfrac{32\sin 84^\circ}{\sin 19^\circ} = \underline{97.75}$

$\dfrac{b}{\sin\beta} = \dfrac{c}{\sin\gamma}$ $\quad\therefore b = \dfrac{32\sin 77^\circ}{\sin 19^\circ} = \underline{95.77}$

4. $\gamma = 180^\circ - (68^\circ + 42^\circ) = \underline{70^\circ}$

$\dfrac{c}{\sin\gamma} = \dfrac{a}{\sin\alpha}$ $\quad\therefore a = \dfrac{760 \times 0.927\ 18}{0.939\ 69} = \underline{750}$

$\dfrac{b}{\sin\beta} = \dfrac{a}{\sin\alpha}$ $\quad\therefore b = \dfrac{a\sin\beta}{\sin\alpha}$ from values for a and γ, $\underline{b = 541}$

Problems 6, 8, and 10 are solved in a similar manner.

6. $a = 2.1$ $\quad c = 4.54$ $\quad \gamma = 114^\circ$

8. $b = 218$ $\quad c = 692$ $\quad \alpha = 56.5^\circ$

10. $a = 256$ $\quad b = 240$ $\quad \gamma = 138.5^\circ$

12.

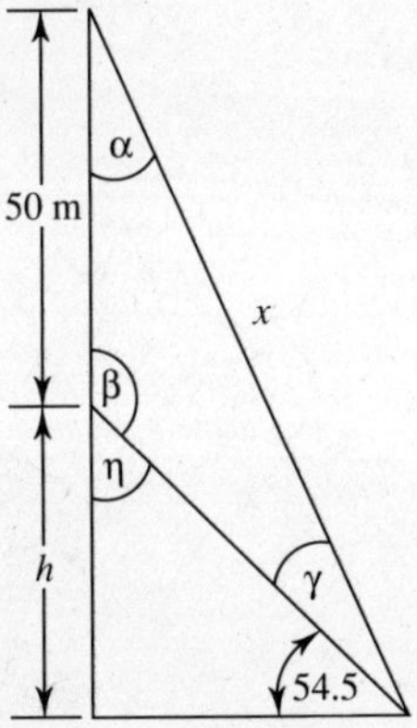

$\alpha = 13.5^\circ$, $\beta = 144.4^\circ$, $\gamma = 22.1^\circ$, $\eta = \alpha + \gamma$

$$\frac{50}{\sin 22.1^\circ} = \frac{x}{\sin 144.4^\circ}$$

$$\therefore x = \frac{50\sin 144.4^\circ}{\sin 22.1^\circ} = \underline{77.7\text{ m}}$$

$$\sin(54.5^\circ + \gamma) = \frac{h + 50}{x}$$

$$\therefore x\sin 76.5^\circ - 50 = h$$

$$77.7(0.972\ 37) - 50 = h$$

$$\underline{\therefore h = 25.5\text{ m}}$$

PROBLEMS 27-3

cosine laws: $a^2 = b^2 + c^2 - 2bc \cos \alpha$

$b^2 = a^2 + c^2 - 2ac \cos \beta$

$c^2 = a^2 + b^2 - 2ab \cos \gamma$

Sample solutions:

2. $c^2 = (474)^2 + (791)^2 - 2(474)(791) \cos 77°$

$c^2 = 850\ 357 - 168\ 684$

$c = \underline{825.6}$

$$\cos \alpha = \frac{b^2 + c^2 - a^2}{2bc} = \frac{(791)^2 + (825.6)^2 - (474)^2}{2(791)(825.6)}$$

$\cos \alpha = 0.828\ 89$

$\therefore \alpha = \underline{34.01°}$

$\beta = 180° - (34.01 + 77°)$ $\therefore \underline{\beta = 68.99°}$

4. $b^2 = (2.6)^2 + (8.45)^2 - (2)(2.6)(8.45)\cos 48.8°$

$b^2 = 78.1625 - 28.9428$

$\underline{b = 7.02}$

$$\alpha = \arccos \frac{b^2 + c^2 - a^2}{2bc} = \arccos 0.960\ 26$$

$\underline{\therefore \alpha = 16.2°}$ $\gamma = 180° - (16.2° + 48.8°) = \underline{115°}$

Problems 6 and 8 are solved in a similar manner.

6. $\underline{a = 0.0091}$, $\underline{\beta = 64.8°}$, $\underline{\gamma = 110.2°}$

8. $\underline{\alpha = 0°}$, $\underline{\beta = 0°}$, $\underline{\gamma = 180°}$, $\cos \gamma = -1$.

$$\text{Since: } \cos \alpha = \frac{(4000)^2 + (6000)^2 - (2000)^2}{2(4000)(6000)} = 1$$

$\underline{\therefore \arccos 1 = \alpha \text{ and } \alpha = 0°}$ (If $\beta = 0°$, γ must be $180°$)

10. Use cosine laws as in Problems 2, 4, 6, and 8.

$\underline{\alpha = 22.3°}$ $\underline{\beta = 27.1°}$ $\underline{\gamma = 130.5°}$

12.

$$A_1 + A_2 + A_3 + A_4 = A_T$$

$$A_1 = A_3 \text{ and } A_2 = A_4$$

$$\therefore A_T = 2A_1 + 2A_2$$

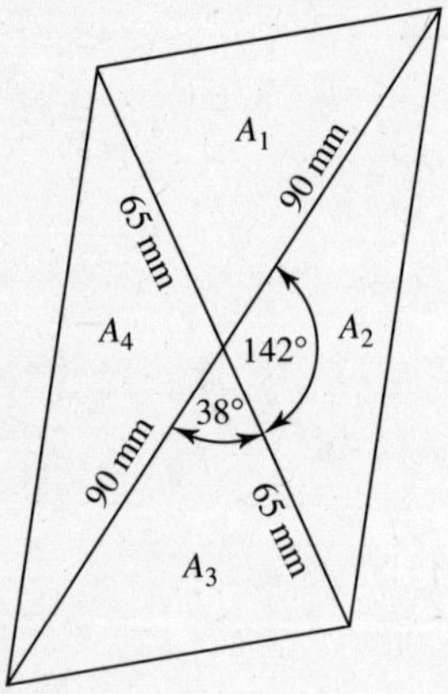

$$A_1 = \frac{1}{2}(65)(90)\sin 38^\circ$$

$$A_2 = \frac{1}{2}(65)(90)\sin 142^\circ$$

$$\sin 142^\circ = \sin(180 - 142^\circ) = \sin 38^\circ$$

$$\therefore A_T = \not{2}\left(\frac{1}{\not{2}}\right)(65)(90)\ \sin 38^\circ + \not{2}\left(\frac{1}{\not{2}}\right)(65)(90)\ \sin 38^\circ$$

$$A_T = (5850 + 5850)\ \sin 38^\circ$$

$$= 11\ 700(0.615\ 66)$$

$$\underline{A_T = 7203.239\ \text{mm}^2}$$

The other method of solution using cosine laws is left as an exercise. Hint: solve one long side and one short side of parallelogram.

PROBLEMS 27-4

2. $(\sin 45^\circ \cos\theta - \cos 45^\circ \sin\theta) - (\cos\theta\cos 45^\circ - \sin 45^\circ \sin\theta)$

$= 0.707\cos\theta - 0.707\sin\theta - 0.707\cos\theta + 0.707\sin\theta$

$\underline{= 0}$

4. $(\sin\theta\cos 30^\circ - \cos\theta\sin 30^\circ) - (\cos\theta\cos 45^\circ + \sin\theta\sin 45^\circ)$

$= 0.866\ 03\sin\theta - 0.5\cos\theta - 0.707\ 11\cos\theta - 0.707\ 11\sin\theta$

$\underline{= 0.158\ 92\sin\theta - 1.207\ 11\cos\theta}$

6. $\sin\theta = \frac{3}{5}$, $\quad\therefore \cos\theta = \frac{4}{5}$, $\sin\phi = \frac{5}{12}$, $\quad\therefore \cos\phi = \frac{10.91}{12}$

Given: $\sin\theta\cos\phi - \cos\theta\sin\phi - \cos\phi\cos\theta - \sin\phi\sin\theta$

Substituting values: $\frac{32.7}{60} - \frac{20}{60} - \frac{43.6}{60} - \frac{15}{60}$

$= -\frac{45.9}{60} = \underline{-0.765}$

PROBLEMS 28-1

The following calculations are made for the purpose(s) of (a) verifying vector diagrams below and (b) providing a calculator exercise using rectangular ⟶ polar interconversions.

2. Add:

$$130\angle 65^\circ = 54.94 + j117.8$$
$$80\angle 235^\circ = -45.9 - j65.5$$
$$160\angle 310^\circ = 103 - j122.6$$
$$\text{sum} = 112 - j70.3$$

Polar form = $132\angle 328^\circ$

4.

$$70\angle 70^\circ = 23.94 + j65.78$$
$$90\angle 115^\circ = -38.04 + j81.57$$
$$110\angle 250^\circ = -37.6 - j103.4$$
$$140\angle 335^\circ = 126.9 - j59.2$$
$$\text{sum} = 75.18 - j15.21$$

Polar form = $76.7\angle 348.6^\circ$

2.

Vector Diagram #1

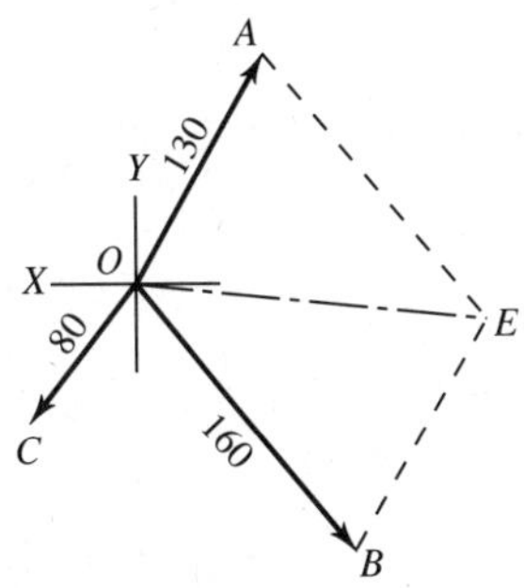

Vector Diagram #2

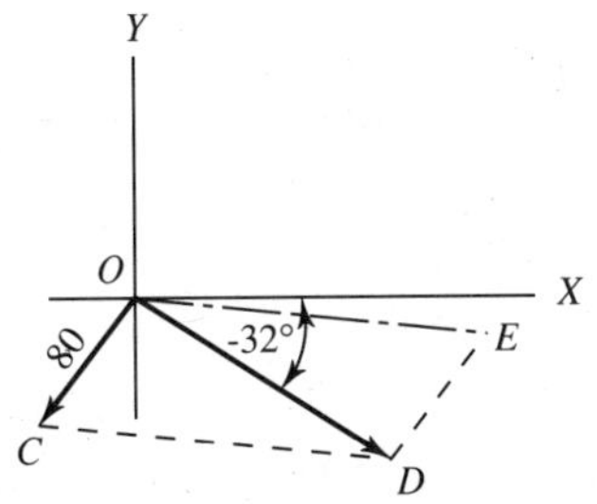

RESULTANT OD = $132\angle 328^\circ$

4.

Vector Diagram #1

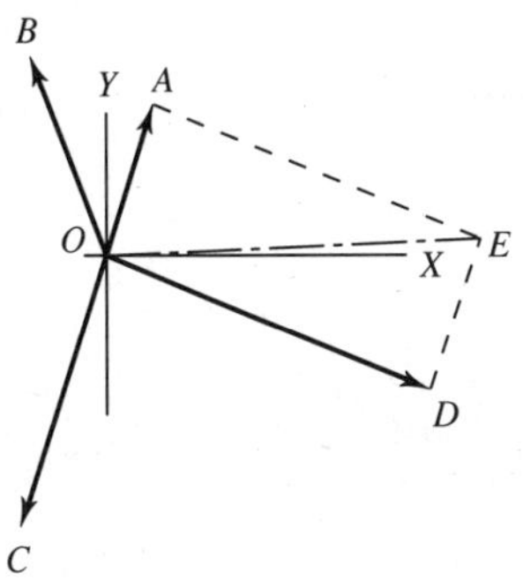

Vector Diagram #2

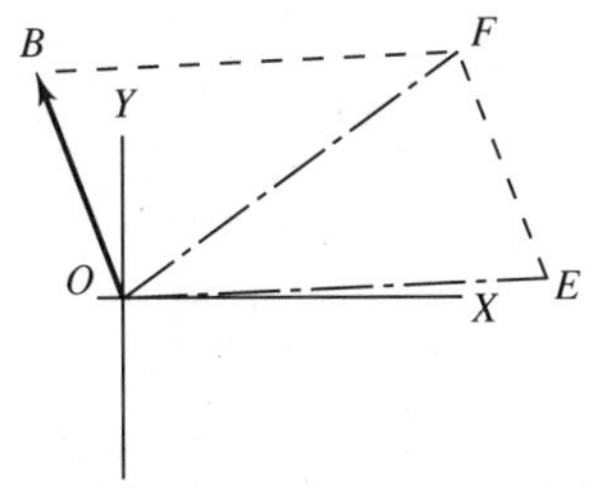

RESULTANT OG = $76.7\angle 348.58^\circ$

Note: The diagrams for Q2 & Q4 have been "separated" to show the steps necessary for solution; redraw the diagrams for Q4, this time on one vector diagram.

Vector Diagram #3

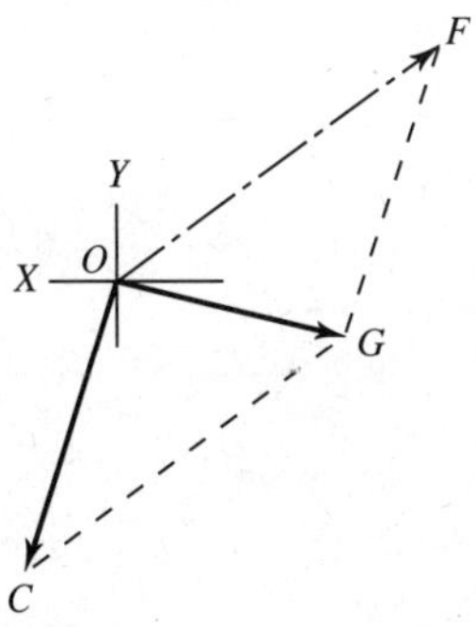

None of the diagrams shown are drawn to any "true" scale.

PROBLEMS 28-2

2. $x = 55.9$, $y = 38.6$. Proof: since $\cos 34.6^\circ = \frac{x}{68}$ and $\sin 34.6^\circ = \frac{y}{68}$,

then $x = 68 \cos 34.6^\circ = \underline{55.97}$ and $y = 68 \sin 34.6^\circ = \underline{38.61}$

4. $x = 1800 \cos 120^\circ = \underline{-900}$ and $y = 1800 \sin 120^\circ = \underline{1559}$

6. $x = 0.987 \cos 295.5^\circ = \underline{0.425}$ and $y = 0.987 \sin 295.5^\circ = \underline{-0.891}$

8. $27.8275 \cos 90^\circ = \underline{x = 0}$ and $27.8275 \sin 90^\circ = y = \underline{27.8275}$

10. $1600 \cos 270^\circ = \underline{x = 0}$ $\quad y = 1600 \sin 270^\circ = \underline{-1600}$

12. 1950 km/h = 541.7 m/s

541.7 cos 82° = horizontal velocity

(541.7)(0.139 17) = 75.4 m/s

14. $250\ N = \sqrt{(155N)^2 + x}$ $\quad \therefore x = \sqrt{[(250)(N)]^2 - (155N)^2}$

$\therefore x = 196N$

components = $196N$ and $155N$

*NOTE: diagrams are left as an exercise.

PROBLEMS 28-3

2. $\tan \theta = \frac{1.97}{12.4}$ $\quad \therefore \theta = \arctan 0.158\ 87 = 9.03^\circ$

$\sin \theta = \frac{1.97}{Z}$, $Z = \frac{1.97}{\sin 9.03^\circ} = \underline{12.6\angle 9.03^\circ}$

4. $\arctan \frac{153}{45.4} = \theta = 73.5^\circ$ $\quad \therefore Z = \frac{153}{\sin 73.5^\circ} = \underline{160\angle 73.5^\circ}$

6. $\theta = 0^\circ$ and $\underline{Z = 864\angle 0^\circ}$

8. $5.27\angle 180^\circ = -5.27\angle 0^\circ$ $\quad \therefore \theta = \arctan \frac{6.0}{-5.27} = \underline{131.3^\circ}$

$Z = \frac{6.0}{\sin 131.3^\circ} = \underline{7.98\angle 131.3^\circ}$ (II Quadrant Solution) $7.986\angle 131.3^\circ$

10. $323\angle 270^\circ = -323\angle 90^\circ$

$\tan \theta = \frac{-323}{323} = -1$ $\quad \arctan -1 = \underline{\theta = -45^\circ}$

$Z = \frac{323}{\sin -45^\circ} = \underline{457\angle -45^\circ}$

12. $37\angle 90^\circ + 34.4\angle 90^\circ = 71.4\angle 90^\circ$ $\quad \therefore \tan \theta = \frac{71.4}{84.2}$, $\underline{\theta = 40.3^\circ}$

$Z = \frac{84.2}{\cos 40.3^\circ} = \underline{110\angle 40.3^\circ}$

14. $270^\circ = -90^\circ$ and $180^\circ = -1\angle 0^\circ$

$167\angle 270^\circ + 81.3\angle 90^\circ = 85.7\angle -90^\circ$

$143.8\angle 180^\circ + 252\angle 0^\circ = 108.2\angle 0^\circ$

$\theta = \arctan(85.7 \div 108.2) \qquad \therefore \ \underline{\theta = 321.6^\circ}$

$$Z = \frac{108.2}{\cos 321.6^\circ} = \underline{138\angle 321.6^\circ} \qquad \left(\text{Show that } Z = \frac{-85.7}{\sin 321.6^\circ}\right)$$

PROBLEMS 28-4

2. $647\angle 51.5^\circ = 402.77 + j506.35$

$215\angle 135^\circ = -152.03 + j152.03$

$250.74 + j658.38 = \underline{705\angle 69.2^\circ}$

4. $7.65\angle 17.8^\circ = 7.28\angle 0^\circ + 2.34\angle 90^\circ$

$4.34\angle 137.5^\circ = 3.2\angle 180^\circ + 2.93\angle 90^\circ$

$\text{sum} = 4.08\angle 0^\circ + 5.27\angle 90^\circ \qquad \therefore \ \theta = \arctan \frac{5.27}{4.08} = 52.3^\circ$

and $Z = 6.67\angle 52.3^\circ$

PROBLEMS 29-2

2. $\frac{3600 \ \cancel{\text{rev}}}{1 \ \cancel{\text{min}}} \times \frac{1 \ \cancel{\text{min}}}{60 \text{ s}} \times \frac{2\pi^r}{1 \ \cancel{\text{rev}}} = \underline{120\pi \text{ r/s}}$

$\frac{3600}{60} \times 360^\circ = 21\,600^\circ/\text{s}$

4. (a) $\frac{1200 \ \cancel{\text{rev}}}{30 \ \cancel{\text{s}}} \times \frac{60 \ \cancel{\text{s}}}{\text{min}} \times \frac{2\pi^r}{\cancel{\text{rev}}} = \underline{4800\pi^r/\text{min}}$

(b) $\frac{1200}{30} \times 60 \times 360 = 864\,000^\circ/\text{min}$

6. (a) $\sin 0.002\pi = \sin 0.0063^r = \sin 0.361^\circ = 0.0063$

(b) $\sin 0.2\pi = \sin 0.628^r = \sin 36^\circ = 0.5878$

(c) $\sin \pi = \sin \pi^r = \sin 180^\circ = 0.000$

(d) $\sin (0.95)(2\pi) = \sin 5.969^r = \sin 342^\circ = -0.3090$

PROBLEMS 29-3

2. $y = r\sin(2\pi ft - 15^\circ)$ $\therefore y = 170\sin(377t - 15^\circ)$

(a) Amplitude = 170 (b) Angular velocity = $2\pi f$ = 377

(c) Frequency = $\dfrac{377t}{2\pi t}$ = 60 Hz

(d) Period = $\dfrac{1}{f} = \dfrac{1}{60}$ s (e) 15° lag

	(a)	(b)	(c)	(d)	(e)
4.	I_{max}	2513	400 Hz	2.5 ms	22° lag
6.	I_{cmax}	1000π	500 Hz	0.002 s	37° lead

14. (b) $y = 175\sin(24\pi t - 40^\circ)$ mm

(c) $y = 175\sin(162^\circ - 40^\circ) = \underline{148.4 \text{ mm}}$

(d) y @ $t = 0.833$ s

$$y = 175\sin[(2\pi \times 12 \times 0.833)^r - 40^\circ]$$

$$y = -115.8\angle{-41.4^\circ} \text{ mm}$$

$$\frac{-115.8}{x} = \tan -41.4^\circ \qquad \therefore x = \frac{-115.8}{\tan -41.4^\circ} = \underline{131.2 \text{ mm}}$$

(e) Radians after 2.5 s = $(2\pi \times 12 \times 2.5)^r = \underline{60\pi^r}$

PROBLEMS 30-1

2. $18.4 = V_{max}\sin 24^\circ$ $\therefore V_{max} = \dfrac{18.4}{\sin 24^\circ} = \underline{45.2 \text{ V}}$

4. $i_{inst} = I_{max}\sin(2\pi ft)^r$

(a) $i = 750 \times 10^{-3}\sin 26^\circ = 329$ mA

(b) $i = 750 \times 10^{-3}\sin 341^\circ = -244$ mA

(c) $i = 750 \times 10^{-3}\sin 210^\circ = -375$ mA

(d) $i = 750 \times 10^{-3}\sin 297^\circ = -668$ mA

(e) $i = 750 \times 10^{-3}\sin 162^\circ = 232$ mA

6. $-22 = V_{max}\sin 289^\circ$ $\therefore V_{max} = \dfrac{-22}{\sin 289^\circ} = \underline{23.3 \text{ V}}$

@ 142° $v = 23.3\sin 142^\circ = 23.3(0.615\ 66)$ V

$$\therefore \underline{v = 14.3 \text{ V}}$$

8. $I_{max} = 365$ mA, 80% $I_{max} = 292$ mA

when $I_{max} = 365 \text{ mA} \angle\theta = 90^\circ$, 0.8 I_{max} occurs twice

when $I_{max} = 365 \sin\theta$ $\qquad \sin\theta = 1$

$292 = 365 \sin\theta$ mA $\qquad \sin\theta = \dfrac{292 \times 10^{-3}}{365 \times 10^{-3}}$

arcsin 0.799 68 = $\underline{\theta = 53.1^\circ}$ I Quadrant

and when I_{max} is 292 (80%), $\theta = 180^\circ - 53.1^\circ = \underline{126.9^\circ}$ II Quadrant

10. $v = 156 \sin(90^\circ + 92^\circ)$ V $\quad \therefore v = -5.44$ V

PROBLEMS 30-2

2. (a) $f = \dfrac{PS}{60} = \dfrac{6 \times 1200}{60}$ (12 poles = 6 pairs of poles) = 120 Hz

(b) $v = 170 \sin 240\pi t$ V

4. $f = \dfrac{PS}{60}$ $\quad \therefore P = \dfrac{f60}{S} = \dfrac{800 \times 60}{4000}$ = 12 pairs of poles

(a) Number of poles = 24 $\qquad$ (b) $v = 163 \sin 1600\pi t$ V

(c) $v = 163 \sin\left(1600\pi \times 500 \times 10^{-6} \times \dfrac{180}{\pi}\right)^\circ$ V

$= 163 \sin 144^\circ$ $\quad \therefore v = 95.8$ V

6. $i = 84.6 \sin 377t$ mA, $\quad 2\pi ft = 377t$ $\quad \therefore f = 60$ Hz

8. $\sin(2\pi ft) = \dfrac{v}{V_{max}}$, $\arcsin \dfrac{v}{V_{max}} = 2\pi ft$, $\sin\theta = \dfrac{346}{400} = 0.865\ 00$;

$\theta = 59.882\ 71^\circ = 1.045\ 15^r$, $f = 4.3$ MHz, $2\pi ft = 1.045\ 15^r$

$\therefore$ (a) $t = \dfrac{1.045\ 15}{2\pi(4.3 \times 10^6)} = \underline{38.68 \text{ ns}}$

(b) $i_{max} = 2.22\ \mu$A

10. $v = 3.62 \times 3.62 \times 10^{-4} \sin(2.86\pi \times 10^6)t$ V

PROBLEMS 30-3

2. $i_{av} = 120 \times 10^{-3}$ A $\quad \therefore I_{max} = I_{av} \times \dfrac{\pi}{2} = 120 \times 10^{-3} \times \dfrac{\pi}{2} = \underline{188.5 \text{ mA}}$

4. $I_{max} = 173\ \mu$A $\quad \therefore i_{av} = I_{max} \times 0.637 = \underline{110\ \mu\text{A}}$

6. $rms = 117$ V $\quad \therefore V_{max} = 117 \times \sqrt{2} = \underline{165 \text{ V}}$

8. $V_{av} = 125$ V, $V_{max} = 196$ V, $rms = 139$ V

10. 33.8 A rms, $I_{max} = 48$ A, $I_{av} = 30.5$ A

PROBLEMS 30-4

2. (a) $700\ \text{A} = 990\ \text{A}_{pk}$ $\qquad \therefore i = 990 \sin(157t + 22^\circ)\ \text{A}$

(b)

$$465 = 990 \sin(157t + 22^\circ)$$
$$0.4697 = \sin(157t + 22^\circ)$$
$$\arcsin 0.4697 = 157t + 22^\circ$$
$$28^\circ = 157t + 22^\circ \qquad \therefore 6^\circ = 157t$$

6° of voltage waveform when $i = 465$ A

4. $i = 990 \sin(350^\circ + 22^\circ)\ \text{A} \qquad \therefore i = 990 \sin 372^\circ \qquad \therefore \underline{i = 206\ \text{A}}$

6. From Prob. 5, $\qquad i = I_{max} \sin(800\pi t - 20^\circ)\ \text{A}$

When $800\pi t^r = 17^\circ$ $\qquad i = 40 \sin(17^\circ - 20^\circ)\ \text{A}$

$$i = 40(-0.052\ 34)$$
$$i = \underline{-2.09\ \text{A}}$$

8. From Prob. 7: I lags V by 9.2°

$$\therefore i = 42.4 \sin(377t - 9.2^\circ)$$
$$\frac{-39.3}{42.4} = \sin(377t - 9.2^\circ)$$
$$\arcsin -0.9268 = 377t - 9.2^\circ$$
$$-67.95^\circ + 9.2^\circ = 377t \qquad \therefore 377t = -58.75^\circ$$
$$v = 170 \sin 377t$$
$$= 170 \sin -58.75^\circ \qquad \therefore v = -145\ \text{V}$$

10. From Prob. 9: $\theta = 49^\circ$ lead or lag

(a) $i = 42.4 \sin(2513t \pm 49^\circ)\ \text{A}$ $\qquad$ (42.4 A = 30 A *rms*)

(b) $-325 = 325 \sin(2513t)$ $\qquad \therefore \sin(2513t) = -1$

$\arcsin -1 = -90^\circ$ $\qquad$ (325 V = 230 *rms*)

$i = 42.4 \sin(-90^\circ \pm 49^\circ)$ $\qquad$ ($\sin -139^\circ \equiv \sin -41^\circ$)

$i = 42.4(-0.656\ 06)$ $\qquad = -0.656\ 06$

$\underline{i = -27.8\ \text{A}}$

PROBLEMS 31-1

2. $99.6 - j85 \qquad \theta = \arctan \dfrac{-85}{99.6} \qquad \theta = -40.5^\circ$

$$Z = \frac{99.6}{\cos -40.5^\circ} = 131 \qquad \therefore \underline{Z = 131\angle -40.5^\circ}$$

4. $5.62 - j40 = 40.4\angle -82^\circ$ (or $40.4\angle 278^\circ$)

6. $-316 - j435 = 538\angle 234^\circ$ (or $538\angle -126^\circ$)

8. $14.9 - j13.6 = \underline{20.2\angle -42.4^\circ}$ *NOTE: scientific calculator with R → P function can verify these answers. Use P → R key(s) to reconvert answers.

10. $-62.1 - j85.8 = \underline{106\angle 234.1^\circ}$

12. $0 - j100 = \underline{100\angle -90^\circ}$

PROBLEMS 31-2

2.
$$\begin{array}{r} 12 + j14 \\ \underline{22 + j17} \\ 264 + j308 \\ \underline{j204 + j^2 238} \\ 264 + j512 + j^2 238 \end{array}$$

$(j^2 = -1)$ $\therefore 26 + j512$

or: $12 + j14 = \underline{18.44\angle 49.4^\circ}$

$22 + j17 = \underline{27.8\angle 37.7^\circ}$

Multiply magnitudes and add angles to produce:

polar ⟶ $= \underline{513\angle 87.1^\circ}$

4. $155\ 122.4 - j11\ 250 = \underline{1.56 \times 10^5 \angle -4.15^\circ}$

6. $-39.6 - j129.6 = \underline{135.5\angle 253^\circ}$

8. $$\frac{7 - j5}{10 - j14} \times \frac{10 + j14}{10 + j14} = \frac{70 + j48 - j^2 70}{100 - j^2 196} = 0.473 + j0.162$$

$$= \underline{0.5\angle 18.9^\circ}$$

or $$\frac{7 - j5}{10 - j14} = \frac{8.6\angle -35.53^\circ}{17.2\angle -54.46^\circ}$$

Divide magnitudes (8.6 ÷ 17.2) and subtract the angles to produce:
⟶ $0.5\angle 18.9^\circ$ (as before)

10. $$\frac{12 + j9}{144 - j^2 81} = \frac{12 + j9}{225} = 0.0533 + j0.04 = \underline{0.0667\angle 36.9^\circ}$$

PROBLEMS 31-3

2. $31[\cos 53.7^\circ + j \sin 53.7^\circ] + [137(\cos -53.7^\circ + j \sin -53.7^\circ)]$

$= 31(0.5920 + j0.8059) + 137(0.5920 - j0.8059)$

$= 99.5 - j85.4$ in polar ⟶ $\underline{131\angle -40.6^\circ}$

4. $P \longrightarrow R = (16.965 - j17.815) + (-11.38 - j22.15)$

sum $= 5.58 - j40$ polar ⟶ $\underline{40.4\angle -82^\circ}$

6. $777(\cos - 129^\circ + j \sin 129^\circ) + 241(\cos 44.3^\circ + j \sin 44.3^\circ)$

$= (-489 - j604) + (172 + j168)$

sum $= -316 - j436$ ⟶ polar $= \underline{538\angle 234^\circ}$

8. $9(\cos -22.2^\circ + j \sin -22.2^\circ) - 12.1(\cos 57.4^\circ + j \sin 57.4^\circ)$

$= (8.33 - j3.4) - (6.519 + j10.194)$

$= 1.81 - j13.6$ ⟶ polar $= \underline{13.7\angle -82.4^\circ}$

10. $85(\cos -27.15^\circ + j \sin -27.15^\circ) - 145(\cos 18.91^\circ + j \sin 18.91^\circ)$

$= (75.6 - j38.79) - (137.2 + j46.99)$

$= -61.5 - j85.8 \longrightarrow$ polar $= 106\underline{/234^\circ}$

12. $10.64(\cos -53^\circ + j \sin -53^\circ) - 22.35(\cos 62.5^\circ + j \sin 62.5^\circ)$

$= (6.4 - j8.5) - (10.3 + j19.8)$

$= -3.92 - j28.3 \longrightarrow$ polar $= 28.6\underline{/262.1^\circ}$

PROBLEMS 31-4

2. $546\underline{/78.2^\circ} = (21.4 \times 25.5)\underline{/52.6^\circ + 25.6^\circ}$

$546(\cos 78.2^\circ + j \sin 78.2^\circ) = 112 + j534 = 546\underline{/78.2^\circ}$

4. $(183.3 \times 3.26)\underline{/11^\circ + (-11^\circ)} = 598\underline{/0^\circ} = 598 + j0$

6. $(9.5 \times 8.26)\underline{/-71.6^\circ - 7.6^\circ} = 78.5\underline{/-79.2^\circ}$

$78.5(\cos - 79.2^\circ + j \sin -79.2^\circ) = 14.7 - j77.1$

8. $\frac{92.3}{81}\underline{/-12.5^\circ + 64.6^\circ} = 1.14\underline{/52.1^\circ}$

$1.14(\cos 52.1^\circ + j \sin 52.1^\circ) = 0.7 + j0.899$

10. $\frac{5}{30}\underline{/20^\circ - (-51.9^\circ)} = 0.1667\underline{/71.90^\circ}$

$0.1667(\cos 71.9^\circ + j \sin 71.9^\circ) = 0.0518 + j0.1584$

12. $\frac{1.87}{3.54}\underline{/-180^\circ + 180^\circ} = 0.528\underline{/0^\circ} = 0.528 + j0$

14. $\sqrt{1024}\underline{/-\frac{17^\circ}{2}} = \pm 32\underline{/-8.5^\circ}$

16. $(0.31)^2\underline{/2 \times -60^\circ} = 0.0961\underline{/-120^\circ}$

18. $\sqrt[3]{1331}\underline{/\frac{17.4^\circ}{3}} = 11.0\underline{/5.8^\circ}$ or $(10.94 + j1.112)$

20. $2^5\underline{/5 \times -16^\circ} = 32\underline{/-80^\circ}$

PROBLEMS 32-1

2. $v = V_{max} \sin \omega t$, $V_{max} = 550 \times \sqrt{2} = 778$ V, $\omega t = 2 \times \pi \times 60 \times t = 377t$

(a) $v = 778 \sin 377t$ V

(b) $i = I_{max} \sin 377t$ A $\quad I_{max} = (6.6)\sqrt{2} = 9.33$ A

$\therefore i = 9.33 \sin 377t$ A

(c) $V_{R_3} = \underline{55\text{ V}}$ (Solution for Problem 2(c) on p. 101)

Problem 32-1 Number 2(c), continued.

$I_4R_4 = 6.6 \times 75 = 495$ V

$V = IR_4 + I_3R_3$ Since $I_1R_1 = I_2R_2 = I_3R_3$

$V - I_4R_4 = I_3R_3$ $\therefore V_{R_3} = 550 - 495$ and $I_3R_3 = \underline{55\ \text{V}}$

(d) $I_{R_1} = \frac{55}{39} = 1.41$ A, $I_{R_3} = \frac{55}{56} = 0.982$ A

$\therefore I_{R_2} = 6.6 - (1.41 + 0.982) = 4.21$ A

$\therefore P_{R_2} = 4.21 \times 55 = \underline{231\ \text{W}}$

(e) $I_{R_1} = \underline{1.41\ \text{A}}$ (solved in part (d))

(f) $v = 778 \sin 377t$, when $v = 36.5$ V, $i = ?$

$\frac{36.5}{778} = \sin 377t$ $\therefore i = \frac{36.5}{778} \times 9.33 = \underline{438\ \text{mA}}$

4. $\frac{V^2}{R} = P$ $\therefore V = \sqrt{RP} = \sqrt{600 \times 800 \times 10^{-3}}$

$V = \sqrt{480} = 21.91$ V $\therefore V_{max} = 21.91 \times \sqrt{2} = \underline{31\ \text{V}}$

PROBLEMS 32-2

Formulae for Problems 2, 4, and 6 $X_L = 2\pi f L$, $L = \frac{X_L}{2\pi f}$, $f = \frac{X_L}{2\pi L}$

2. $X_L = 75.4\ \Omega$ **4.** $L = 300$ mH **6.** $f = 183$ MHz

8. $v = 141 \sin 1600\pi t$ $I_{max} = \frac{V_{max}}{X_L} = 419$ mA

$\therefore \underline{i = 419 \sin (1600\pi t - 90^\circ)\ \text{mA}}$

10. $314t = 2\pi f t$ $\therefore f = 50$ Hz, $X_L = 2\pi f L = 518\ \Omega$

$I_{max} = \frac{311}{518} = 600$ mA $\therefore \underline{i = 600 \sin (314t - 90^\circ)\ \text{mA}}$

12. $i = 600 \sin (314t - 90^\circ)$ mA, then $314t = 300^\circ$, and $(314t - 90^\circ) = 210^\circ$

$v = 311 \sin 314t$ $\therefore v = 311 \sin 300^\circ = 311(-0.8660)$ $\therefore v = -269$ V

14. X_L varies directly as L.

PROBLEMS 32-3

2. $X_C = \frac{1}{2\pi f C} = \frac{1}{2\pi \times 10^3 \times 50 \times 10^{-6}} = \underline{3.18\ \Omega}$

4. $\underline{X_C = 0.265\ \Omega}$

6. $\underline{X_C = 75.4\ \Omega}$

8. When $f = 12$ kHz, $X_C = 5.3$ MΩ

$i_C = \frac{V}{X_C} = \frac{125}{5.3} \times 10^{-6} = \underline{23.6\ \mu A}$

10. $V = IX_C$

$= \frac{452 \times 10^{-3}}{2\pi \times 60 \times 5 \times 10^{-6}} = 240$ V

12. $i = I_{max} \sin(2\pi f t + 90^\circ)$ mA, $I_{max} = 452 \times \sqrt{2}$ and $2\pi f t = 230^\circ$

$\therefore i = (452 \times \sqrt{2})(\sin 230^\circ + 90^\circ) = \underline{-411\ \text{mA}}$

14. $\frac{1}{C_t} = \frac{1}{C_1} + \frac{1}{C_2}$ $\quad \therefore \underline{C_t = 18.2\ \text{pF}}$

16. (a) X_C is halved

(b) X_C is reduced to one third

(c) X_C is doubled

18. (a) $v = (220 \times \sqrt{2})(\sin 120\pi t)$

$v = 311 \sin 377t$

(b) $i = 138 \sin(377t + 90^\circ)$ mA

(c) $V_{C_1} = IX_{C_1} = \frac{97.6 \times 10^{-3}}{2\pi \times 60 \times 2.5 \times 10^{-6}} = \underline{103.6\ V}$

(d) $V_{C_2} = V_G - V_{C_1} = 220 - 103.6 = \underline{116.4\ V}$

20. From $X_C = \frac{1}{2\pi f C}$ (a) $X_C = \underline{2980\ \Omega}$ (b) $X_C = \underline{4081\ \Omega}$

PROBLEMS 32-4

2. (a) $X_L = 2\pi f L = 2\pi \times 8 \times 10^6 \times 500 \times 10^{-6} = \underline{25.1\ k\Omega}$

(b) $Z = R + jX_L = 3.3 \times 10^3 + j25.1 \times 10^3 = \underline{25.3\ k\Omega}$

[No angles shown]

(c) $i_t = \frac{V_G}{Z} = \frac{500}{25.3} \times 10^{-3} = \underline{19.7\ \text{mA}}$

(d) $i_t Z = i_t(R + jX_L)$ current phase angle is $\underline{82.5^\circ\ \text{lag}}$

(e) $v_R = i_t R = 3.3 \times 10^3 \times 19.7 \times 10^{-3} = \underline{65\ V}$

(f) $v_L = i_t X_L = 25.1 \times 10^3 \times 19.7 \times 10^{-3} = \underline{496\ V}$

4. (a) $X_L = 2\pi \times 400 \times 2.8 = \underline{7037\ \Omega}$

$$Z = R + jX_L = 1210 + j7037 = \underline{7140\ \Omega\angle 80.2^\circ}$$

(b) $i = \dfrac{V}{Z} = \dfrac{50}{7140} = \underline{7.00\ \text{mA}}$ (to three significant figures)

[No angles shown]

6. $X_C = \dfrac{1}{2\pi \times 120 \times 22 \times 10^{-6}} = 60.3\ \Omega$

$Z = 330 - j60.3 = 335.5\ \Omega$ (angle = -10.4°)

$i_C = \dfrac{V_G}{Z} = \dfrac{120}{335.5} = \underline{358\ \text{mA}}$

8. $Z = (6.7 + 0.33) \times 10^3 - j121 = 7031\ \Omega = \underline{7.03\ \text{k}\Omega}$ (X_C is negligible)

$i = \dfrac{120}{7.03} \times 10^{-3} = 17.07 = \underline{17.1\ \text{mA}}$

10.

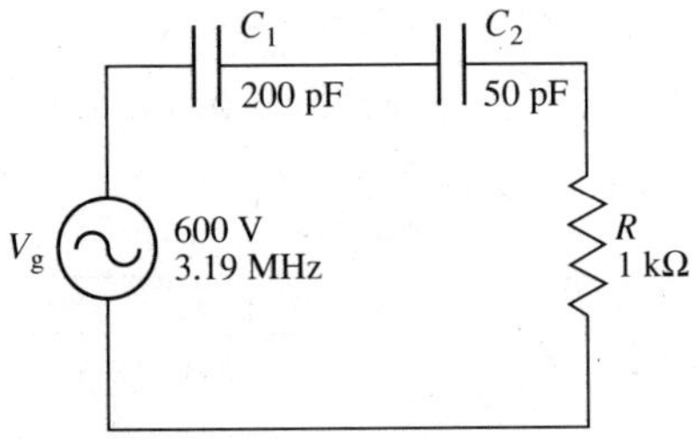

$C_T = \dfrac{C_1 C_2}{C_1 + C_2} = \dfrac{10\ 000}{250} = 40$ pF

From Prob. 9: $f = 3.19$ MHz

$V_g = 600\angle -51.3^\circ$ V

$X_{C_1} = 250\angle -90^\circ\ \Omega$, $X_{C_2} = 997.8\angle -90^\circ\ \Omega$

$X_{C_t} = X_{C_1} + X_{C_2} = 1247.8\angle -90^\circ\ \Omega$

(a) $I\angle\theta^\circ = \dfrac{V_g\angle\theta^\circ}{Z\angle\theta^\circ} = \dfrac{600\angle -51.3^\circ}{1599\angle -51.3^\circ} = 375\angle 0^\circ$ mA; (b) $V_R = \underline{IR = 375\ \text{V}}$

(c) $V_{C_1} = I(X_{C_1}) = \underline{94\angle -90^\circ\ \text{V}}$; (d) $V_{C_2} = I(X_{C_2}) = \underline{374\angle -90^\circ\ \text{V}}$

(e) $Z\angle\theta^\circ = R + jX_{C_t} = 1000 + (-j1247.8) = \underline{1599\angle -51.3^\circ\ \Omega}$

PROBLEMS 32-5

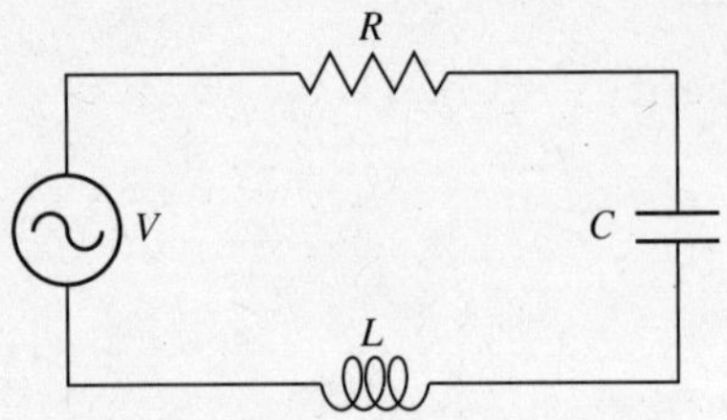

diagram for problems 2, 4, 6, 8, and 10

2. (sample solution)

$V = 450$ V, $f = 1$ kHz, $R = 67\ \Omega$

$L = 5$ mH, $C = 50\ \mu$F

$X_L = 2\pi \times 5 = 31.42\ \Omega$, $X_C = \frac{1}{\omega C} = 3.183\ \Omega$

(a) $Z = R + j(X_L - X_C)$

$= 67 + j(31.42 - 3.183)$

$= \underline{72.7\angle 22.9^\circ\ \Omega}$

(b) $I = \frac{V}{Z} = \frac{450}{72.7} = \underline{6.19\text{ A}}$

(c) $i = I_{max} \sin (2\pi f t^r - 22.9^\circ)$ A $\qquad I_{max} = 6.19 \times \sqrt{2}$ A, $2\pi f t = 6280t$

$= 8.75 \sin (6280t - 22.9^\circ)$ A

(d) $PF = \cos \theta = \cos 22.9^\circ = 0.92$ or $\underline{92\%}$

(e) Power expended $= IV \cos \theta$ W $= 6.19 \times 450 \times 0.92 = \underline{2.57\text{ kW}}$

Solutions for problems 4-10 are not shown; the tabulated answers on this page were calculated using diagram and method above.

	Z	I	i	PF%	P
2.	$72.7\angle 22.9^\circ\ \Omega$	6.19 A	$i = 8.75 \sin (6280t - 22.9^\circ)$ A	92	2.57 kW
4.	$6.3\text{k}\angle 85.4^\circ\ \Omega$	135 mA	$i = 191 \sin (2512t - 85.4^\circ)$ mA	7.9	9.11 W
6.	$333\angle 7.32^\circ\ \Omega$	3.01 A	$i = 4.26 \sin (5.03 \times 10^{10}t - 7.32^\circ)$ A	99.2	2.99 kW
8.	$28.9\angle -21.1^\circ\ \Omega$	69.1 mA	$i = 97.8 \sin (62\ 800t + 21.1^\circ)$ mA	93.3	129 mW
10.	$50.5\angle -8.5^\circ\ \Omega$	2.18 A	$i = 3.08 \sin (377t + 8.5^\circ)$ A	98.9	237 W

12.

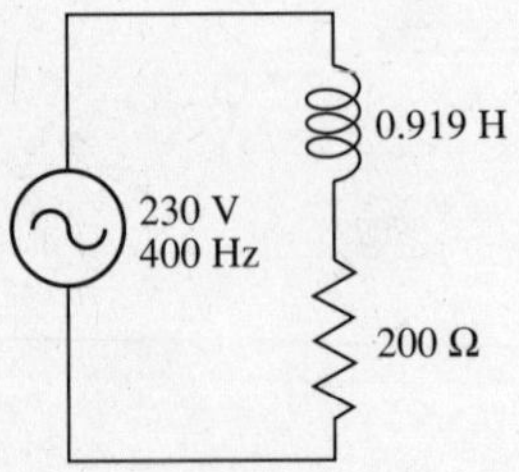

$X_L = 800\pi \times 0.919 = 2310\angle 90^\circ\ \Omega$

$Z = 200 + j2310 = 2319\angle 85.05^\circ\ \Omega$

$I = \frac{V}{Z} = \frac{230}{2319\angle 85.05^\circ} = 99.2\angle -85.05^\circ$ mA

Power $= I^2R = (0.0992)^2(200) = 1.97$ W

Power $= IV \cos 85.05^\circ$ (angle should *not* be "rounded off") $= (0.0992)(230)(0.086\ 29) = 1.9688$ W

$\underline{\text{Power} = 1.97\text{ W}}$

14. $I^2R = P \quad \therefore R = 23\ \Omega$

$P = IV\cos\theta \quad \therefore \theta = \frac{2300}{4400} = 0.522\ 73 \quad \therefore \theta = 58.5^\circ$

$Z = \frac{V}{I} = \frac{440}{10} = 44\ \Omega$ since: $Z = R + jX_L$, $44 = 23 + jX_L$

$X_L = Z\sin\theta$ and $R = Z\cos\theta = 23\ \Omega$ and $X_L = 44(0.8525) = 37.5\ \Omega$

Output sees: $23 + j37.5\ \Omega = (44\angle 58.5^\circ)\ \Omega$

16. $v = v_R + v_X$ (no phase angle for instantaneous values)

PROBLEMS 32-6

2. $Q = \frac{X_L}{R_S} = \frac{2\pi f L}{R_S} = \frac{\omega L}{R_S} = \frac{2\pi \times 15 \times 10^3 \times 265 \times 10^{-3}}{1100} = 22.7$

4. @ resonance $X_L = X_C$

@ $f_r, C = \frac{1}{4\pi^2 f^2 L} = \frac{1}{4\pi^2 \times (10^4)^2 \times 265 \times 10^{-3}} = 955.86$ pF

$\therefore$ with 500 pF in circuit, (955.86 - 500) pF = 455.86 pF in parallel.

@ $f_r, L = \frac{1}{4\pi^2 f^2 C} = 506.6$ mH $\quad \therefore$ (506.6 - 265) mH = 241.6 mH in series

6. @ resonance (a) $R + j0$; below f_r (b) $R - jX$; above f_r (c) $R + jX$

8. @ 1 MHz, $X_L = X_C$, from Prob. 7, $X_L = 1502\ \Omega$, $X_C = \frac{1}{\omega C}$, $C = \frac{1}{X_C}$

$C = 106$ pF, $\therefore$ 6 pF should be added in parallel with original 100 pF.

(a) $Q = \frac{X_C}{R} = \frac{1502}{15} = 100$ (c) $v_C = v_L = I_g X_L = I_g X_C$. $I_g = \frac{0.3\text{ mV}}{15} = 20\ \mu\text{A}$

(b) $v_R = v_g = 0.3$ mV $\quad \therefore v_L = 20 \times 10^{-6} \times 1502 = 30$ mV @ $+90^\circ$

since $v_L - v_C = 0$ V $\quad$ or $v_c = Qv_g = (100 \times 0.3\text{ mV}) = 30$ mV @ -90°

(d) At resonance the phase angle between current and voltage must be zero.

PROBLEMS 33-1

2.

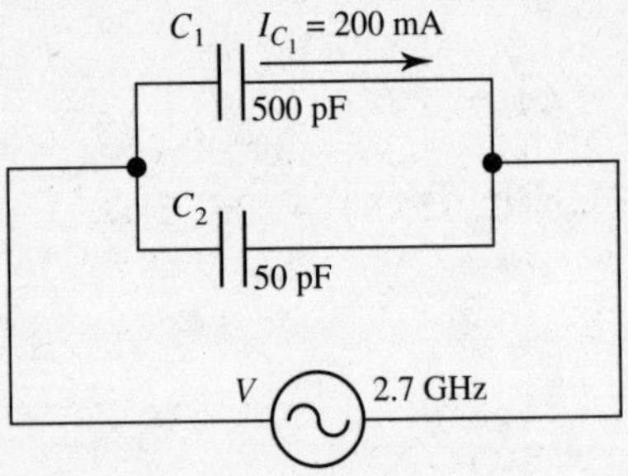

$$X_{C_1} = \frac{1}{\omega 500 \times 10^{-12}} = 0.1179\ \Omega$$

$$X_{C_2} = 10 \geq X_{C_1} = 1.179\ \Omega \text{ and } V_{C_1} = V_{C_2} = V$$

$$\therefore V = I_{C_1} \times X_{C_1} = 200 \times 10^{-3} \times 0.1179$$

$$= \underline{23.58 \text{ mV}}$$

$$I_{C_2} = \frac{V}{X_{C_2}} = \frac{23.58 \times 10^{-3}}{1.179} = \underline{20 \text{ mA}}$$

$$\text{Note: } I_{C_2} = \frac{I_{C_1}}{10} = 20 \text{ mA}$$

4. $X_L = \omega L = 377 \times 2 = 754\ \Omega$, $R = 200\ \Omega$

$$X_C = \frac{1}{\omega C} = \frac{1}{377 \times 5 \times 10^{-6}} = 530\ \Omega$$

$I_R = 1.1 + j0$

$I_L = 0 \quad - j0.292$

$I_C = 0 \quad + j0.415$

(a) $I_T = 1.1 + j0.123 \text{ A} = \underline{1.11\angle 6.38^\circ \text{ A}}$

(b) Power $= IV \cos\theta = 1.11 \times 220 \times 0.9938 = 243$ W

(c) $Z = \dfrac{V}{I} = \dfrac{220}{1.11\angle 6.38^\circ} = 198\angle -6.38^\circ\ \Omega$

$R = Z \cos\theta = 198 \cos -6.38^\circ = 197\ \Omega$

$X = Z \sin -6.38^\circ = 22\angle -90^\circ\ \Omega$ or $-j22\ \Omega$

Equivalent series circuit $= 197 - j22\ \Omega$

(d) $PF = \cos\theta = 0.994$ or 99.4% leading

(e) $i = 1.57 \sin(377t + 6.38^\circ)$ A

6.

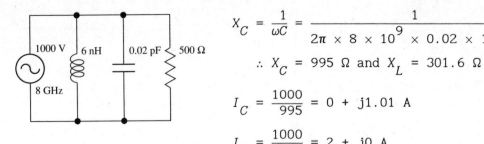

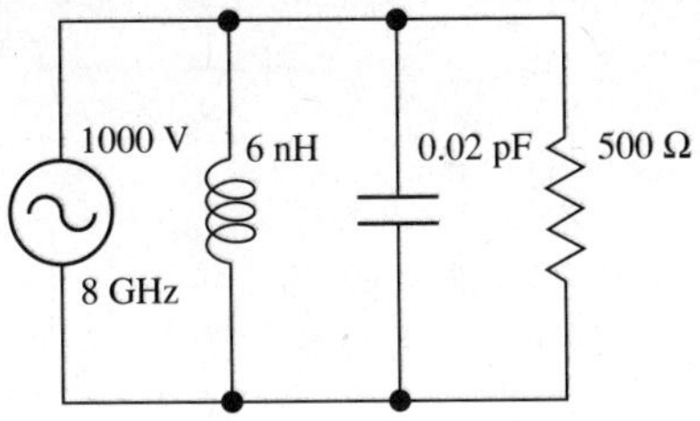

$$X_L = \omega L = 2\pi \times 8 \times 10^9 \times 6 \times 10^{-9} = 301.6\ \Omega$$

$$X_C = \frac{1}{\omega C} = \frac{1}{2\pi \times 8 \times 10^9 \times 0.02 \times 10^{-12}}$$

$$\therefore X_C = 995\ \Omega \text{ and } X_L = 301.6\ \Omega$$

$$I_C = \frac{1000}{995} = 0 + j1.01\ \text{A}$$

$$I_R = \frac{1000}{500} = 2 + j0\ \text{A}$$

$$I_L = \frac{1000}{301.6} = 0 - j3.32\ \text{A}$$

(a) $I_T = I_R + I_L + I_C = 2 + j(I_C - I_L) = 2 - j2.31\ \text{A} = 3.1\angle{-49.1^\circ}\ \text{A}$

(b) For $PF = 1$, total X_C MUST EQUAL total X_L (resonance)

Let C_P be required capacitor for $X_L = X_C$.

$$301.6 = \frac{1}{2\pi \times 8 \times 10^9 \times C_P}$$

$$\therefore C_P = \frac{1}{16\pi \times 10^9 \times 301.6} = \underline{0.066\ \text{pF}}$$

Additional capacitance = 0.066 - 0.02 = 0.046 pF

PROBLEMS 33-2

2. $Z_a = 148.5\angle{42.2^\circ} = 110 + j99.75$

$Z_b = 145\angle{-12.7^\circ} = 141.5 - j31.88$

$Z_a + Z_b = 251.5 + j67.87 = 260.5\angle{15.1^\circ}\ \Omega$

$$Z_t = \frac{Z_a Z_b}{Z_a + Z_b} = \frac{(148.5\angle{42.2^\circ})(145\angle{-12.7^\circ})}{260.5\angle{15.1^\circ}} = \underline{82.7\angle{14.4^\circ}\ \Omega}$$

4. $Z_a = 276 - j180 = 330\angle{-33.1^\circ}\ \Omega$

$Z_b = 117 - j18.6 = 118.5\angle{-9.03^\circ}\ \Omega$

$Z_a + Z_b = 393 - j198.6 = 440\angle{-26.8^\circ}\ \Omega$ $\quad\quad \therefore Z_t = \dfrac{Z_a Z_b}{Z_a + Z_b}$

$$Z_t = \frac{(330\angle{-33.1^\circ})(118.5\angle{-9.03^\circ})}{440\angle{-26.8^\circ}} = \underline{88.9\angle{-15.3^\circ}\ \Omega}$$

6. $Z_1 = 347 + j73.8 = 355\angle 12^\circ\ \Omega$

$Z_2 = 0 - j100 = 100\angle -90^\circ\ \Omega$

$Z_1 + Z_2 = 347 - j26.2 = 348\angle -4.32^\circ\ \Omega$

Solved as in prior problems, then:

$$Z_t = \frac{Z_1 Z_2}{Z_1 + Z_2} \qquad \therefore Z_T = 102\angle -73.7^\circ\ \Omega$$

8. $$Z_t = \frac{Z_1 Z_2}{Z_1 + Z_2} \qquad \therefore Z_1 = \frac{Z_t Z_2}{Z_2 - Z_t}$$

$Z_t = 53.5\angle -42.4^\circ\ \Omega$

$Z_2 = 168\angle 27^\circ\ \Omega$

$$\therefore Z_1 = \frac{(53.5\angle -42.4^\circ)(168\angle 27^\circ)}{168\angle 27^\circ - 53.5\angle -42.4^\circ}\ \Omega$$

$$Z_1 = \frac{8988\angle -15.4^\circ}{157.4\angle 45.6^\circ} = 57.1\angle -61^\circ\ \Omega$$

10. $Z_1 = 78.5 - j35 = 85.95\angle -24.03^\circ$

$Z_2 = 33.6 + j48 = 58.59\angle 55.01^\circ$

$Z_1 + Z_2 = 112.1 + j13 = 112.85\angle 6.61^\circ$

$$Z_t = Z_s + \frac{Z_1 Z_2}{Z_1 + Z_2} = 9.4 + j6.6 + \frac{(85.95\angle -24.03^\circ)(58.59\angle 55.01^\circ)}{112.85\angle 6.61^\circ}$$

$Z_t = 9.4 + j6.6 + 44.62\angle 24.37^\circ$

$Z_t = 9.4 + j6.6 + 40.64 + j18.41$

$\therefore Z_t = 50.04 + j25.01 = 55.9\angle 26.6^\circ\ \Omega$

12. $Z_1 = 57.2\angle -61^\circ = 27.73 - j50.03$

$Z_2 = 168\angle 27^\circ = 149.69 + j76.27$

$Z_1 + Z_2 = 179.35\angle 8.41^\circ = 177.42 + j26.24$

$$Z_t = 5 + j3.9 + \frac{Z_1 Z_2}{Z_1 + Z_2} = 5 + j3.9 + 53.58\angle -42.41^\circ$$

$$= 5 + j3.9 + 39.56 - j36.14$$

$\therefore Z_t = 44.56 - j32.24 = 55\angle -35.9^\circ\ \Omega$

14. $$I_s = \frac{-j\omega MV}{Z_pZ_s + (\omega M)^2} \qquad Z_pZ_s = (6 + j8)(20 + j12) = 24 + j231.96$$

$$\therefore I_s = \frac{(15)(20)\angle -90^\circ}{24 + j231.96 + 15^2} = \frac{300\angle -90^\circ}{249 + j231.96}$$

$$I_s = \frac{300\angle -90^\circ}{340\angle 42.97^\circ} \qquad \therefore \underline{I_s = 882\angle -133^\circ \text{ mA}}$$

PROBLEMS 33-3

2.

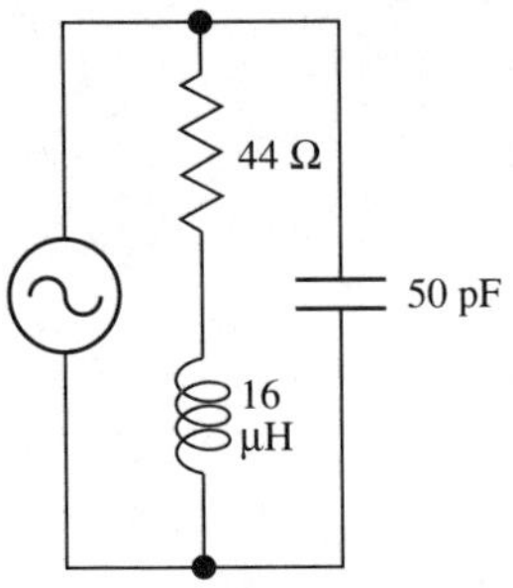

(a) $$f_o = \frac{1}{2\pi}\sqrt{\frac{1}{LC} - \frac{R^2}{L^2}}$$

$$= \frac{1}{2\pi}\sqrt{\frac{1}{12 \times 60 \times 10^{-18}} - \frac{44^2}{(12 \times 10^{-6})^2}}$$

$$= \frac{1}{2\pi}\sqrt{1388.9 \times 10^{12} - 13.44 \times 10^{12}}$$

$$\therefore f_o = \underline{5.90 \text{ MHz}}$$

(b)* $$f_o = \frac{1}{2\pi\sqrt{LC}} = \frac{1}{2\pi} \times \frac{1}{\sqrt{12 \times 60 \times 10^{-18}}} = \underline{5.931 \text{ MHz}}$$

(c) $$Q = \frac{\omega L}{R} = \frac{2\pi \times 5.931 \times 10^6 \times 12 \times 10^{-6}}{44} = \underline{10.16}$$

*Note: $f_o = \frac{1}{2\pi\sqrt{LC}}$ is only an approximation; A $Q >> 10$ is implied. Usually effective in 98% of situations.

4. $P = \frac{V^2}{Z_t}$ From Prob. 3: $Z_t = \frac{L}{CR} = \omega LQ = 2\pi f LQ$

$$\therefore Z_t = 2\pi f \times 10 \times 10^{-3} \times 800$$

$$f = \frac{1}{2\pi\sqrt{LC}} = 112.54 \text{ kHz}$$

$$Z_t = \frac{2\pi \times 10 \times 10^{-1} \times 8}{2\pi\sqrt{10 \times 10^{-3} \times 200 \times 10^{-12}}} = \underline{5.656 \text{ M}\Omega}$$

$$\therefore P = \frac{(600)^2}{5.656} \times 10^{-6} = \underline{63.6 \text{ mW}}$$

6.

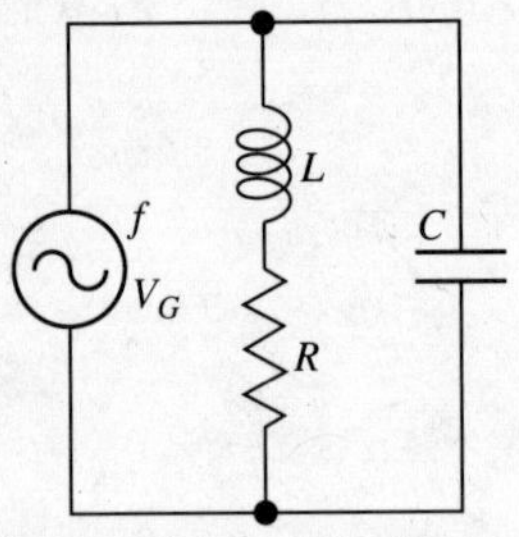

diagram for problems 6 and 8

(a) Given: f_r = 999 kHz, Q = 90, C = 254 pF

$$X_L = X_C = \frac{1}{2\pi \times 999 \times 10^3 \times 254 \times 10^{-12}}$$

$$X_C = 627.2\ \Omega$$

$$\therefore L = \frac{X_L}{2\pi f} = \frac{627.2}{2\pi \times 999 \times 10^3} = 99.9\ \mu\text{H}$$

(a) L = 99.9 μH [Use 100 μH (within 1%)]

(b) $Q = \frac{X_L}{R_S} = \frac{627.2}{R_S} = 90$ $\qquad \therefore \underline{R_S = 6.97\ \Omega}$

(c) $Z_t = \omega LQ = 627.2 \times 90 = \underline{56.4\ \text{k}\Omega}$

8. Data: V_G = 20 V, f = 499 kHz, L = 99.9 μH (not 100 μH)

R = 6.97 Ω and C = 254 pF.

$$X_L = \omega L = 2\pi \times 499 \times 10^3 \times 99.9 \times 10^{-6} = 313.22\ \Omega$$

$$Z_L = 6.97 + j313.22 = 313.3\angle 88.72^\circ\ \Omega$$

$$X_C = \frac{1}{\omega C} = \frac{1}{2\pi \times 499 \times 10^3 \times 254 \times 10^{-12}} = 1255.7\ \Omega$$

$$I_L = \frac{V_G}{Z_L} = \frac{20\angle 0^\circ}{313.3\angle 88.72^\circ} = 63.8\angle -88.72^\circ\ \text{mA}$$

$$I_C = \frac{V_G}{X_C} = \frac{20\angle 0^\circ}{1255.7\angle -90^\circ} = 15.9\angle 90^\circ\ \text{mA}$$

$$I_t = I_L + I_C = (1.426 - j63.822) + j15.9\ \text{mA}$$

$$= 1.426 - j47.922$$

$$\therefore I_t = 47.9\angle -88.3^\circ\ \text{mA}$$

Power = $VI \cos\theta$ = 20 × 47.9 × 0.029 74 = $\underline{28.5\ \text{mW}}$

$PF = \cos\theta$ = 0.029 74 or 2.97% lagging.*

*This answer is identical to one using 100 μH coil.

10. Q = 100, f_r = 7.496 MHz, Z_t = 65.9 kΩ. Find L.

$$Z_t = 2\pi f_r LQ \qquad \therefore L = \frac{65\ 900}{2\pi \times 7.496 \times 10^6 \times 100} = \underline{14\ \mu\text{h}}$$

12. $I_t = 18.9$ mA, $f_r = 1.5$ MHz, $V = 1$ kV, $Q = 99.7$. Find C.

$$Z_t = \frac{V}{I_t} = \frac{1000}{18.9} \times 10^3 = 53\ \text{k}\Omega. \quad Z_t = \omega LQ \quad L = \frac{Z_t}{\omega Q} = 56\ \mu\text{H}$$

$$\omega L = \frac{1}{\omega C} \qquad \therefore C = \frac{1}{\omega^2 L} \text{ but } \frac{1}{L} = \frac{\omega Q}{Z_t} \qquad \therefore C = \frac{Q}{\omega Z_t}$$

$$\therefore C = \frac{99.7}{2\pi \times 1.5 \times 10^6 \times 53 \times 10^3} \cong \underline{200\ \text{pF}} \quad (199.59\ \text{pF exact})$$

14. (a) $I_t = 10.3$ mA, (b) $R_s = 39$ kΩ, (c) $L = 8$ mH, $C = 3.14$ pF
(d) $V_R = V_g = 400\angle 0^\circ$ V

PROBLEMS 33-4

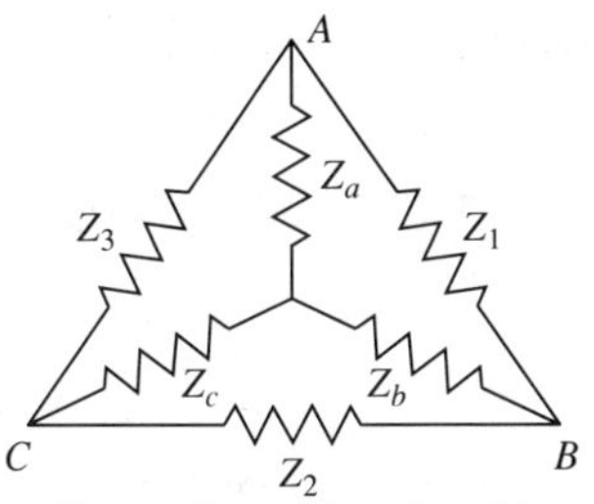

diagram for problems 2 and 4

$$Z_a = \frac{Z_1 Z_3}{Z_1 + Z_2 + Z_3} \quad \text{Eq. (1)}$$

$$Z_b = \frac{Z_1 Z_2}{Z_1 + Z_2 + Z_3} \quad \text{Eq. (2)}$$

$$Z_c = \frac{Z_2 Z_3}{Z_1 + Z_2 + Z_3} \quad \text{Eq. (3)}$$

Note: Do not round off until final answer. Differences in calculator operation may cause different intermediate answers; final answers should agree to three significant figures.

2. $Z_1 + Z_2 + Z_3 = \Sigma Z_\Delta$

$Z_1 = 3 + j4 = 5\angle 53.13^\circ$ Ω

$Z_2 = 12 + j5 = 13\angle 22.62^\circ$ Ω

$Z_3 = 8 - j6 = 10\angle -36.87^\circ$ Ω

$\Sigma Z_\Delta = 23 + j3 = 23.195\angle 7.43^\circ$ Ω

(Try $5\angle 53.1^\circ$ P ⟶ R function)
Did you get 3.002 + j3.998?
$5\angle 53.13^\circ$ is exactly 3 + j4*

From Eq. (1) $Z_a = \dfrac{(5\angle 53.13^\circ)(10\angle -36.87^\circ)}{23.195\angle 7.43^\circ} = \underline{2.156\angle 8.829^\circ\ \Omega\ (2.16\angle 8.83^\circ\ \Omega)}$

Using Eq. (2) and data above: $\underline{Z_b = 2.8\angle 68.3^\circ\ \Omega}$ $(2.802\angle 68.269^\circ)$

Using Eq. (3), $Z_c = 5.605\angle -21.73^\circ = \underline{5.61\angle -21.7^\circ\ \Omega}$

*Solve for Z_a if rounded off values for Z_1, Z_2, and Z_3 are used: $Z_1 = 5\angle 53.1^\circ$ $Z_2 = 13\angle 22.6^\circ$ $Z_3 = 10\angle -36.9^\circ$

Round off ΣZ_Δ to $23.2\angle 7.43^\circ$. Does $Z_a = 2.16\angle 8.8^\circ$?

4. $Z_a = 50.9\angle 86.8^\circ = 2.841 + j50.821\ \Omega$

$Z_b = 62.7\angle -20.2^\circ = 58.844 - j21.650\ \Omega$

$Z_c = 44.5\angle 8.8^\circ = 43.976 + j6.8\ \Omega$

$$Z_1 = \frac{Z_aZ_b + Z_bZ_c + Z_aZ_c}{Z_c} \text{ and } Z_aZ_b + Z_bZ_c + Z_aZ_c = \Sigma Z_Y$$

$$Z_2 = \frac{\Sigma Z_Y}{Z_a} \quad \text{and } Z_3 = \frac{\Sigma Z_Y}{Z_b}$$

$Z_aZ_b = (50.9\angle 86.8^\circ)(62.7\angle -20.2^\circ) = 3191.4\angle 66.6^\circ\ \Omega$

$Z_bZ_c = (62.7\angle -20.2^\circ)(44.5\angle 8.8^\circ) = 2790.2\angle -11.4^\circ\ \Omega$

$Z_aZ_c = (50.9\angle 86.8^\circ)(44.5\angle 8.8^\circ) = 2265.1\angle 95.6^\circ\ \Omega$

Substitute into equations for Z_1, Z_2, and Z_3 and prove:

$\underline{Z_1 = 134.4\angle 42^\circ\ \Omega}$, $\underline{Z_2 = 117.5\angle -36^\circ\ \Omega}$

$\underline{Z_3 = 95.4\angle 71^\circ\ \Omega}$ (solutions rounded off)

6.

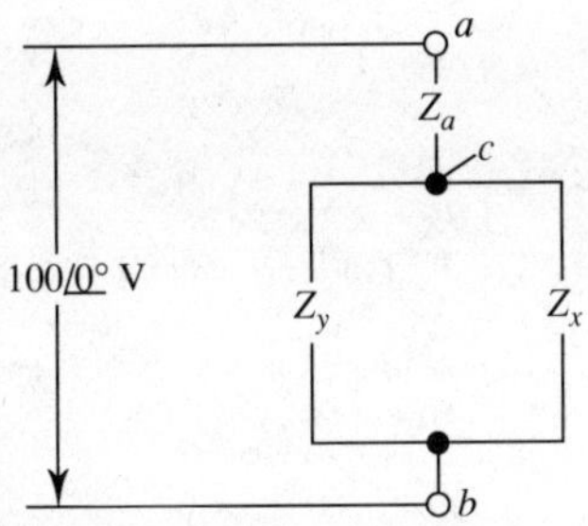

$Z_x = Z_c + Z_4$

$Z_a = 19.35\angle 68.62^\circ\ \Omega$, $Z_y = 103.7\angle 42.6^\circ\ \Omega$

$Z_x = 86.16\angle -54.07^\circ\ \Omega$

From Prob. 5, rounded value for Z_{ab}

$$Z_{ab} = 76.1\angle 2.86^\circ\ \Omega$$

$$I_T = \frac{100\angle 0^\circ}{76.1\angle 2.86^\circ} = 1.314\angle -2.86^\circ \text{ A.}$$

$V_{ac} = 1.314\angle -2.86^\circ \times 19.35\angle 68.62^\circ$

$V_{ac} = 25.43\angle 65.76^\circ$ V

$V_{cb} = 100\angle 0^\circ - V_{ac}$

$= 89.56 - j23.19$

working only:

$100 + j0$
$(-)10.44 + j23.19$
$= 89.56 - j23.19$

$\therefore V_{cb} = 92.51\angle -14.52^\circ$ V

$$\frac{V_{cb}}{Z_x} = I_{Z_x} = I_{Z_4} = \frac{92.51\angle -14.52^\circ}{86.16\angle -54.07^\circ} = 1.074\angle 39.55^\circ \text{ A}$$

$\therefore \underline{I_{Z_4} = 1.07\angle 39.6^\circ \text{ A}}$

8.

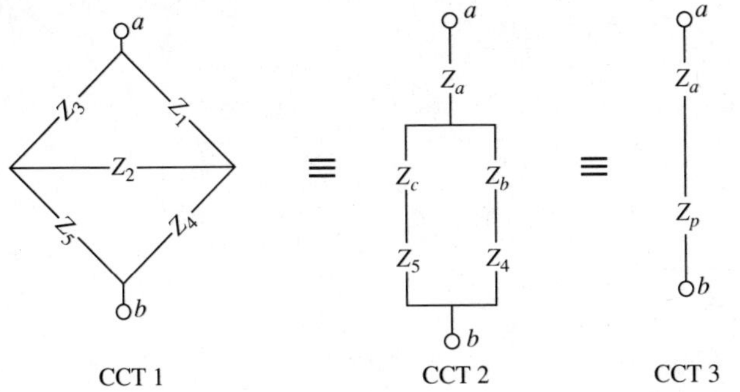

From Prob. 7: $Z_{ab} = 187.4\angle 27.1^\circ\ \Omega$, $\quad Z_a = 152.1\angle 37.73^\circ\ \Omega$

$$I_t = \frac{V}{Z_{ab}} = \frac{440\angle 0^\circ}{187.4\angle 27.1^\circ} = 2.3478\angle -27.1^\circ\ \text{A}$$

$$V_{Z_a} = 2.3478\angle -27.1^\circ \times 152.1\angle 37.73^\circ$$

$$= 357.1\angle 10.63^\circ\ \text{V}$$

$$= 350.97 + j65.87$$

$$V_P = (440 + j0) - (350.97 + j65.87)$$

$$= 89.03 - j65.87$$

$$= 110.75\angle -36.5^\circ\ \text{V}$$

$$I_5 = \frac{V_P}{Z_C + Z_S} = \frac{110.75\angle -36.5^\circ}{196.896\angle -37.3^\circ} = 562.5\angle 0.83^\circ\ \text{mA}$$

$$\therefore I_5 = 563\angle 0.83^\circ\ \text{mA}$$

Data for Prob. 8: (Using rounded off calculations)

$$Z_1 = 102 + j190 = 215\angle 61.8^\circ\ \Omega$$

$$Z_2 = 134 - j33 = 138\angle -13.8^\circ\ \Omega$$

$$Z_3 = 380 - j210 = 434\angle -28.93^\circ\ \Omega$$

$$\Sigma Z_\Delta = 618.3\angle -4.92^\circ \Rightarrow 616 - j53\ \Omega \text{ and } Z_C = \frac{Z_2 Z_3}{\Sigma Z_\Delta}\ \Omega$$

$$\therefore Z_C = 96.9\angle -37.81^\circ\ \Omega \qquad Z_S\text{: given } (100\angle -36.9^\circ)$$

10. Use CCT 1 and CCT 2 from solution for Prob. 8.

From Prob. 9: $V_{Z_4} = 0.848\angle{-35.05^\circ} \times 50\angle{-53.1^\circ}$

$= 42.4\angle{-88.15^\circ}$ V

$$I_5 = \frac{V_P}{Z_C + Z_5} = \frac{50.273\angle{-36.44^\circ}}{197.128\angle{-37.33^\circ}} = 255\angle{0.89^\circ} \text{ mA}$$

$\therefore V_{Z_4} = 0.255\angle{0.89^\circ} \times 100\angle{-36.9^\circ} = 25.5\angle{-36^\circ}$ V

$V_{Z_2} = V_{Z_4} - V_{Z_5} = (1.368\,75 - j42.3779) - (20.6284 - j14.9903)$

$= -19.259\,65 - j27.3876 = 33.482\angle{-125^\circ}$ V

$\therefore V_{Z_2} = 33.482\angle{234.88^\circ}$ V

$$I_2 = \frac{V_{Z_2}}{Z_2} = \frac{33.482\angle{234.88^\circ}}{138.186\angle{-13.836^\circ}} = \underline{242\angle{249^\circ} \text{ mA}}$$

12. From Prob. 11: $Z_{ab} = 88.95\angle{-25.024^\circ}\ \Omega$

$$I_t = \frac{100\angle{0^\circ}}{88.95\angle{-25.024^\circ}} = 1.124\angle{25.024^\circ} \text{ A}$$

$V_{Z_a} = 1.124\angle{25.024^\circ} \times 63.27\angle{-44.456^\circ}$

$= 71.1298\angle{-19.432^\circ} = 67.078 - j23.664$ V

$V_P = 100\angle{0^\circ} - V_{Z_a} = 100 + j0$

$67.078 - j23.664$

$40.544\angle{35.708^\circ} \Leftarrow V_P = 39.922 + j23.664$ V

$$I_5 = \frac{V_P}{Z_5 + Z_C} = \frac{40.544\angle{35.708^\circ}}{55.0114\angle{26.811^\circ}} = 737\angle{8.897^\circ} \text{ mA}$$

$\therefore \underline{I_5 = 737\angle{8.9^\circ} \text{ mA}}$

14.

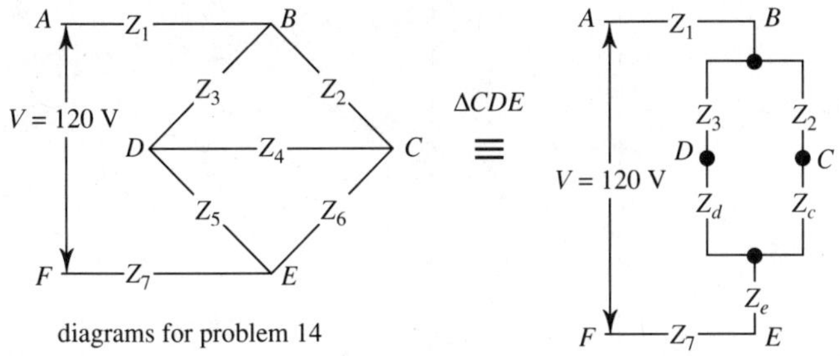

diagrams for problem 14

$Z_1 = 3 + j4 = 5\angle 53.1^\circ\ \Omega,\qquad Z_2 = 37\angle 77.5^\circ = 8.0083 + j36.123\ \Omega$

$Z_3 = 40\angle -80^\circ = 6.946 - j39.392\ \Omega,\qquad Z_4 = 64 - j50 = 81.216\angle -38^\circ\ \Omega$

$Z_5 = 15 + j85 = 86.313\angle 80^\circ\ \Omega,\ Z_6 = 40 - j36 = 53.815\angle -42^\circ\ \Omega$

$Z_7 = 6 - j8 = 10\angle -53.1^\circ\ \Omega$

$I_{Z_7} = I_t = \frac{V}{Z_t},\qquad Z_t = Z_1 + (Z_2 + Z_c)//(Z_3 + Z_d) + Z_e + Z_7$

Solve ΔCDE for values of Z_c, Z_d, and Z_e in Z_Y

$$Z_Y = \frac{Z//Z}{\Sigma Z_\Delta}\qquad \Sigma Z_\Delta = Z_4 + Z_5 + Z_6 = 119\angle -1^\circ$$

$$Z_c = \frac{Z_4Z_6}{\Sigma Z_\Delta} = \frac{(81.216\angle -38^\circ)(Z_6)}{119\angle -1^\circ} = 36.728\angle -79^\circ\ \Omega$$

$$= 7.008 - j36.05$$

$$Z_d = \frac{Z_4Z_5}{119\angle -1^\circ} = 58.907\angle 43^\circ = 43.082 + j40.175\ \Omega$$

$Z_e = 39\angle 39^\circ = 30.334 + j24.564\ \Omega$

$Z_2 + Z_c = 15.0163 - j0.73 = 15.034\angle -2.783^\circ\ \Omega$

$Z_3 + Z_d = 50.028 + j0.783 = 50.034\angle 0.897^\circ\ \Omega$

$$(Z_2 + Z_c)//(Z_3 + Z_d) = \frac{(15.034\angle -2.783^\circ)(50.034\angle 0.897^\circ)}{65.0443 + j0.053}$$

$$= 11.565\angle -1.933^\circ\ \Omega = 11.558 - j0.39\ \Omega$$

$$Z_t = 3 + j4 + 11.558 - j0.390 + 30.334 + j24.564 + 6 - j8$$

$$= 50.892 + j20.174 = 54.745\angle 21.624^\circ\ \Omega$$

$$I_t = \frac{120\angle 0^\circ}{54.745\angle 21.624^\circ} = \underline{2.19\angle -21.6^\circ\ \text{A}}$$

16.

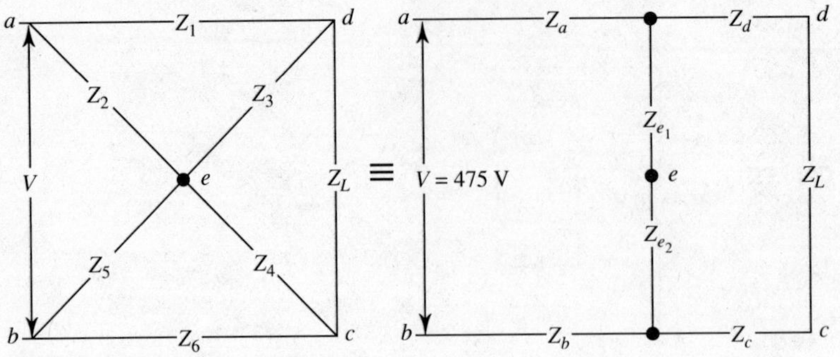

From Prob. 15, solve Δade_1 and Δbce_2.

$Z_a = 378.268\angle 131.75^\circ = -251.893 + j282.199\ \Omega$

$Z_d = 540.206\angle -27.948^\circ = 477.205 - j253.176\ \Omega$

$Z_{e_1} = 650.799\angle -30.448^\circ = 561.049 - j329.794\ \Omega$

$Z_{e_2} = 140.874\angle -85.419^\circ = 11.252 - j140.434\ \Omega$

$Z_b = 198.975\angle 90.681^\circ = -2.365 + j198.961\ \Omega$

$Z_c = 122.287\angle -83.819^\circ = 13.167 - j121.576\ \Omega$

$Z_t = Z_a + Z_b + (Z_{e_1} + Z_{e_2})//(Z_d + Z_L + Z_c) = 279.097\angle 58.13^\circ\ \Omega$

$$I_t = \frac{V}{Z_t} = \frac{475\angle 0^\circ}{279.097\angle 58.13^\circ} = 1.702\angle -58.13^\circ\ \text{A}$$

Solve for I_L: $I_t = I_e + I_L$

$\therefore$ voltage across parallel branch $= Z_P(I_e + I_L)$ V

or $475 - (I_t Z_a + I_t Z_b) = I_t Z_P$

$$I_L = \frac{I_t Z_P}{Z_d + Z_c + Z_L} = \frac{(1.07\angle -58.13^\circ)(469.994\angle -31.3^\circ)}{1228.9\angle -17.76^\circ}$$

$\therefore I_L = 650.93\angle -71.67^\circ$ mA

Extra data for Prob. 16:

$Z_{e_1} + Z_{e_2} = 740.697\angle -39.4^\circ = 572.3 - j470.218\ \Omega$

$Z_c + Z_d + Z_L = 1170.372\angle 0^\circ + 374.751\angle -90^\circ = 1228.9\angle -17.76^\circ$

$(Z_{e_1} + Z_{e_2})//(Z_c + Z_d + Z_L) = Z_P\ \Omega = 401.614 - j244.134$

$Z_a + Z_b + Z_P = Z_t\ \Omega$, $\quad Z_L = 680 + j0 = 680\angle 0^\circ\ \Omega$

18. Data from Prob. 17, diagram for Prob. 16.

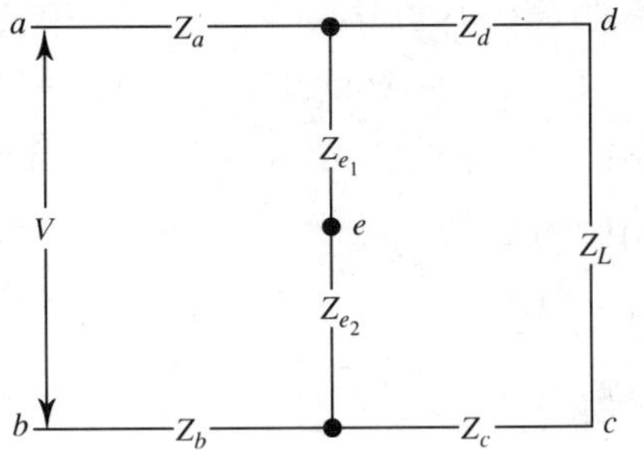

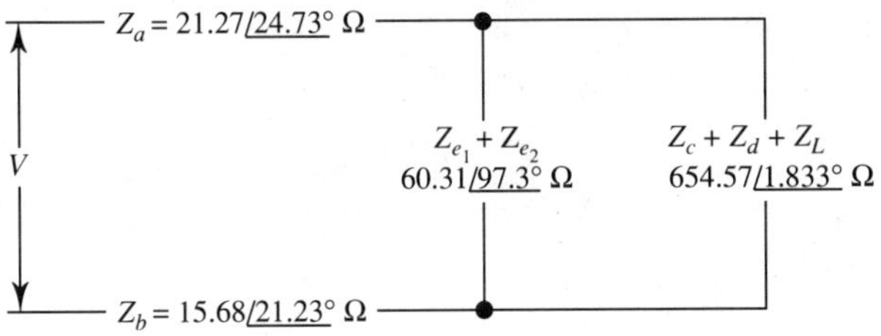

$$Z_a = 21.27\angle 24.73^\circ = 19.32 + j8.899\ \Omega$$

$$Z_b = 15.68\angle 21.23^\circ = 14.61 + j5.68\ \Omega$$

$$Z_c = 26.2\angle 18.15^\circ = 24.88 + j8.156\ \Omega$$

$$Z_d = 32.05\angle 23.51^\circ = 29.39 + j12.78\ \Omega$$

$$Z_{e_1} = 31.03\angle 97.7^\circ = -4.138 + j30.75\ \Omega$$

$$Z_{e_2} = 29.28\angle 96.97^\circ = -3.55 + j29.07\ \Omega$$

$$Z_L = 600\angle 0^\circ = 600 + j0\ \Omega$$

$$Z_t = Z_a + Z_b + (Z_{e_1} + Z_{e_2})//(Z_c + Z_d + Z_L)\ \Omega$$

$$= 21.27\angle 24.73^\circ + 15.68\angle 21.23^\circ + (60.31\angle 97.3^\circ)//(654.57\angle 1.833^\circ)$$

$$= 81.54\angle 67.1^\circ\ \Omega$$

$$V = 135\angle 0^\circ\ \text{V}. \qquad I_t = \frac{V}{Z_t} = 1.656\angle -67.1^\circ\ \text{A}$$

$$Z_S = Z_a + Z_b = 21.27\angle 24.73^\circ + 15.678\angle 21.23^\circ = 36.886\angle 23.25^\circ$$

$$V_S = I_t Z_S = (1.656\angle -67.1^\circ)(36.886\angle 23.25^\circ) = 61.1\angle -43.9^\circ\ \text{V}$$

$Z_P = (Z_{e_1} + Z_{e_2})//(Z_c + Z_d + Z_L)$ and $V_P = I_t Z_P$ or $V - V_S$

$$\therefore V_P = 135\angle 0^\circ - 61.1\angle -43.9^\circ = 135 + j0 - (44.03 - j42.37)$$

$$V_P = 90.97 + j42.37 = 100.36\angle 24.97^\circ\ \text{V}$$

$$I_L = \frac{V_P}{Z_c + Z_d + Z_L} = \frac{100.36\angle 24.9^\circ}{654.57\angle 1.833^\circ} = 0.153\angle 23.1^\circ\ \text{A}$$

Power dissipated by load, $P = I_L^2 R_L$ W

$\therefore P = (0.153)^2 \times 600$ and $\underline{P = 14.05\ \text{W}}$

PROBLEMS 34-1

Note: For numbers less than one, most pocket calculators display the "whole" logarithm as a negative quantity. For example: log 0.200 is shown as -0.698 97. Any calculator answer given will be enclosed in braces { }. Thus, log 0.2 = $\bar{1}$.301 03, {-0.698 97}. Not all problems in this chapter will have calculator answers provided, but all answers may be checked by using a calculator, or five-figure log tables.

2. $\log_{10} 1000 = 3$ 4. $\log_3 81 = 4$ 6. $\log_\varepsilon \varepsilon = 1$ 8. $\log_{10} 10 = 1$

10. $\log_{10} 1 = 0$ 12. $10^5 = 100\,000$ 14. $7^2 = 49$

16. $a^0 = 1$ 18. $25^{0.5} = 5$ 20. $3^{2x} = M$

22. x must equal 6 24. x must equal 5 26. x must equal 4

28. $\log_p 1 = 0$, exponential form; $P^0 = 1$. $\log_7 1 = 0$ $\therefore 7^0 = 1 = P^0$

(Any number raised to a power of 0 is equal to 1.)

30. $\log_3 3 = x$, $3^x = 3 \therefore x = 1$

$\log_3 9 = x$, $3^x = 9 \therefore x = 2$

Similarly: $\log_3 27 = 3$, $\log_3 81 = 4$, $\log_3 243 = 5$

$\log_3 729 = 6$, $\log_3 2187 = 7$

PROBLEMS 34-2

2. 1.861 53 4. -3.138 47 6. 3.000 87

8. -3.999 13 10. 4.518 51

PROBLEMS 34-3

2. $9.304\,649 \times 10^5$ 4. 0.000 093 046 or 9.3046×10^{-5}

6. 7.570 072 8. 757.0072×10^{-6}

10. 0.006 518 3

PROBLEMS 34-4

2. 1.477 **4.** 7.26×10^3 **6.** $P = 11.2$

8. $V = 1.2 \times 10^{-2}$ V **10.** $x = 4.64$ **12.** $x = 2.25$ **14.** 5.40

16. $m^{2.5} = 80 \qquad \therefore 2.5 \log m = \log 80$

$$\log m = \frac{\log 80}{2.5} = \frac{1.903\ 09}{2.5}$$

$$= 0.761\ 236$$

$$\therefore m = 5.77$$

18. $\log \frac{y^2}{4} = 2.2$

$$\frac{y^2}{4} = \text{antilog } 2.2 = 158.489$$

$$\therefore y^2 = (158.489)(4) = 633.95$$

$$y = 25.178\ 51$$

or

$$2 \log y - \log 4 = 2.2$$

$$2 \log y = 2.2 + \log 4$$

$$= 2.2 + 0.602\ 06$$

$$\log y = \frac{2.802\ 06}{2}$$

$$y = \text{antilog } 1.401\ 03$$

$$\therefore y = 25.178\ 51$$

20. $10 \log \frac{P}{1.5} = 32$

$$\log P - \log 1.5 = \frac{32}{10} = 3.2$$

$$\log P = 3.2 + \log 1.5$$

$$= 3.2 + 0.176\ 08 = 3.376\ 09$$

$$P = \text{antilog } 3.376\ 09$$

$$\therefore P = 2.38 \times 10^3$$

22. $\log X^{5.44} - \log X^{2.78} = 1.786$

$$5.44 \log X - 2.78 \log X = 1.786$$

$$2.66 \log X = 1.786$$

$$\log X = \frac{1.786}{2.66} = 0.671\ 43$$

$$\therefore X = \text{antilog } 0.671\ 43 = 4.692\ 76$$

24. $x = \log_6 1296$

$$= \frac{\log_{10} 1296}{\log_{10} 6}$$

$$= \frac{3.112\ 61}{0.778\ 15}$$

$\therefore x = 4$

26. $L_1 = \sqrt[3]{L_2^2}$ (solve for L_2)

$$\log L_1 = \log L_2^{\frac{2}{3}} = \frac{2}{3} \log L_2$$

$$\frac{3}{2} \log L_1 = \log L_2$$

or $\log L_1^{\frac{3}{2}} = \log L_2$

$$\therefore L_2 = L_1^{\frac{3}{2}} = \sqrt{L_1^3}$$

28. $V_g = \frac{2.3T}{11\ 600} \log \frac{I_0}{I_g}$ (solve for I_0)

$$V_g\left(\frac{11\ 600}{2.3T}\right) = \log \frac{I_0}{I_g} = \log I_0 - \log I_g$$

$$5043.5\frac{V_g}{T} + \log I_g = \log I_0$$

$$\therefore I_0 = \left(10^{5043.5\frac{V_g}{T}}\right) I_g$$

30. (a) $V = i_c R \cdot e^{+\left(\frac{t}{RC}\right)}$

(b) $C = \frac{0.4343t}{R(\log V - \log i_c R)}$

(c) $t = 2.3026RC(\log V - \log i_c R)$

32. (a) $V = \frac{I_L R}{1 - e^{-\left(\frac{Rt}{L}\right)}}$

(b) $L = \frac{0.4343Rt}{\log V - \log (V - i_L R)}$

(c) $t = \frac{2.3026\ L}{R}[\log V - \log (V - i_L R)]$

34. (a) $V = \left(\frac{I_P + I_g}{k}\right)^{\frac{2}{3}} - \frac{V_P}{\mu}$

(b) $V_P = \mu\left[-V + \left(\frac{I_P + I_g}{k}\right)^{\frac{2}{3}}\right]$

(c) $\mu = \frac{V_P}{\left(\frac{I_P + I_g}{k}\right)^{\frac{2}{3}} - V}$

36. $i = 1.01$ A

38. 34.7 ms

40. 208 ms

42. 265 wpm

44. 12 ms

46. (a) $1.66 \times 10^{-4} C$

(b) 8 V

48. 296 mA

50. 1.26

PROBLEMS 34-5

2. -247

4. 3.0977×10^{11}

6. -577.927

8. -2.490 74

10. 769.78×10^{-12}

12. 565.642×10^{-9}

14. 312.608×10^{-3}

PROBLEMS 35-1

2. $\sqrt[24]{10^0} = \frac{\log 1}{24} = 1$ = preferred value

0.1 = difference

10% = percentage difference

$\frac{0.5}{1.05} = 0.0476 = \pm 4.76\%$ = tolerance

$10^{\frac{3}{24}} = \frac{3 \log 10}{24} = 1.33$ preferred value = 1.3

difference = 0.2

% difference = 15.04

$\frac{0.1}{1.4} = 0.0714$ ∴ tolerance = ±7.14%

∴ series of R_{24} preferred values:

1.0, 1.1, 1.2, 1.3, 1.5, 1.6, 1.8, 2.0, 2.2, 2.4, 2.7, 3.0, 3.3, 3.6, 3.9, 4.3, 4.7, 5.1, 5.6, 6.2, 6.8, 7.5, 8.2, 9.1, 10.

Maximum % error = ±7.14%

4. R_5 series:

$10^{\frac{x}{5}}$ x	Preferred values	Difference	Tolerance
0	1	60	±23.1%
1	16(1.585)	90	±21.95%
2	25 (2.51)	150	±23.1%
3	40 (3.98)	230	±22.35%
4	63 (6.31)	370	±22.7%
5	100 (10.0)	600	±23.1%

Published tolerance will probably be ±25%

6. (b) 680 Ω, R_6, R_{12}, R_{24}

(d) 470 Ω, R_6, R_{12}, R_{24}

(± 5% resistors usually belong to only one R series) See Table 8 in Appendix.

PROBLEMS 35-2

2. (a) $dB = 20 \log \frac{V_2}{V_1}$

$\therefore dB = 20 \log 32$

$= 30.103$

$\underline{= 30 \text{ dB}}$

(b) $dB = 20 \log 90$

$\underline{= 39 \text{ dB}}$

(c) $dB = 20 \log \frac{1}{120}$

$= 20 \log 0.008\ 33$

$= 20(7.920\ 81 - 10)$

$= 158.4 - 200$

$\underline{= -42 \text{ dB}}$

(d) $20 \log \frac{1}{150}$

$\underline{= -44 \text{ dB}}$

4. $P = \frac{V^2}{R}$ W $\quad \therefore V^2 = (6 \times 10^{-3})(6 \times 10^2) \quad \therefore \sqrt{3.6} = V = \underline{1.9 \text{ V}}$

6. $dB = 20(\log V_2 - \log V_1)$, 12 dBm = 12 dB above 1 mW

$12 \text{ dB} = 10 \log \frac{P_2}{1 \text{ mW}}$

Divide by 10.

$1.2 = \log \frac{P_2}{1 \text{ mW}} \quad$ antilog: $10^{1.2} = \frac{P_2}{1 \text{ mW}} \quad \therefore P_2 = 15.8$ mW

$V = \sqrt{PR} = \sqrt{15.810^{-3} \times 600} = \sqrt{9.5} \quad \underline{\therefore V = 3.08 \text{ mV}}$

8. $V = \sqrt{PR} = \sqrt{12.5 \times 10^{-3} \times 73} = \sqrt{0.913} \quad \therefore V = 9.55$ mV

$I = \frac{V}{R} = \frac{0.955}{73} = 13.08$ mA. Now show that $I = \sqrt{\frac{P}{R}}$ mA also!

10. (a) $5 = 10(\log P_2 - \log 1 \times 10^{-3}) \quad \therefore 10^{0.5} = \frac{P_2}{1 \text{ mW}} \quad \therefore P_2 = 3.16$ mW

Using the same method: $V = \sqrt{PR} = \sqrt{3.16 \times 10^{-3} \times 600} \quad \therefore V = 1.38$ V

(b) $P_2 = 10$ mW, $V = 2.45$ V

(c) $P_2 = 100$ mW, $V = 7.75$ V

(d) $P_2 = 10^{-1}$ mW, $V = 245$ mV

12. $75 \text{ dB} = 20(\log I_2 - \log I_1) \quad \therefore \log \frac{I_2}{I_1} = 3.75$

$\frac{I_2}{I_1} = 10^{3.75} = 5.62 \times 10^3$

14. $50 \text{ dB} = 10 \log \frac{P_2}{P_1} \quad \therefore \log \frac{P_2}{P_1} = 5, \quad \therefore \frac{P_2}{P_1} = 10^5$

16. $dB = 10 \log \frac{3}{10^{-7}} = 10 \log (3 \times 10^7) = 10(7.4771) = 75$ dB

$\underline{\therefore \text{noise is } -75 \text{ dB (below signal)}}$

18. $80\text{ dB} = 20\log\frac{V_2}{V_1}$

$$\log\frac{V_2}{V_1} = 4 \qquad \therefore 10^4 = \frac{V_2}{V_1},\ \text{ratio} = \frac{1}{10^{-4}} = 10^4$$

20. $60 = 10\log\frac{P_2}{P_1} \qquad \log\frac{P_2}{P_1} = 6$

$$\frac{12.5\times10^{-3}}{P_1} = 10^6 \qquad \therefore \underline{P_1 = 12.5\text{ nW}}$$

22. 6 mW ⟶ [−81.5 dB] ⟶ [?] ⟶ 25 W

$$\text{dB} = 10\log\frac{25}{6\times10^{-3}} = 10\log 4.1667\times10^3 = 36.2\text{ dB}$$

Amplification required = 36.2 + 81.5 = 117.7 = <u>118 dB</u>

24. $$\frac{P_2}{P_1} = \frac{\dfrac{V_2^2}{R_2}}{\dfrac{V_1^2}{R_1}} = \left(\frac{V_2^2}{V_1^2}\right)\left(\frac{R_1}{R_2}\right) = (2.92\times10^6)^2\left(\frac{200}{2700}\right)$$

$$\log\frac{P_2}{P_1} = 2(\log 2.92\times10^6) + \log 200 - \log 2700$$

$$= 12.930\ 77 + 2.301\ 03 - 3.431\ 36$$

$$= 11.800\ 43$$

$$\therefore \frac{P_2}{P_1} = 6.315\ 85\times10^{11} = \underline{6.32\times10^{11}}$$

26. @ -5 dB; $-5 = 10 \log \frac{P_2}{30}$ $\quad \therefore -0.5 = \log \frac{P_2}{30}$

$\frac{P_2}{30} = 10^{-0.5}$ $\quad \therefore P_2 = \frac{30}{\sqrt{10}}$ and $\log P_2 = \log 30 - \frac{1}{2} \log 10$

$= 1.477\,12 - 0.5$

$\therefore P_2 = 9.49$ W

@ -10 dB $\frac{P_2}{30} = 10^{-1}$ $\quad \therefore P_2 = 3$ W

@ -15 dB $\frac{P_2}{30} = 10^{-1.5}$ $\quad \therefore P_2 = 0.949$ W

@ -20 dB $\frac{P_2}{30} = 10^{-2}$ $\quad \therefore P_2 = 0.3$ W

@ -25 dB $\frac{P_2}{30} = 10^{-2.5}$ $\quad \therefore P_2 = 0.0949$ W

@ -30 dB $\frac{P_2}{30} = 10^{-3}$ $\quad \therefore P_2 = 0.03$ W

28. $23 = 10 \log \frac{45 \times 10^{-3}}{P_1}$ $\quad \therefore \frac{45 \times 10^{-3}}{P_1} = 10^{2.3}$

$\therefore P_1 = \frac{45 \times 10^{-3}}{10^{2.3}} = \frac{45}{1.9953 \times 10^{2}} = 2.26 \times 10^{-1}$ mW

$\therefore P_1 = 226\ \mu$W

30. (a) $P_{out} = \frac{V^2}{R} = \frac{250^2}{3500} = 17.86$ W

(b) $P_{in} = \frac{0.3^2}{300} = 300\ \mu$W

$\text{dB} = 10 \log \frac{P_{out}}{P_{in}} = 10 \log \left(\frac{17.86}{300} \times 10^6\right)$ $\quad \therefore$ power gain = 47.75 dB

(c) $A_v = \frac{v_{out}}{v_{in}} = \frac{250}{0.3} = 833$

32. $\text{dB} = 10 \log \frac{60 \times 10^{-6}}{P_{in}}$ and $P_{in} = \frac{(9 \times 10^{-6})^2}{80} = 1.0125 \times 10^{-12}$ W

$= 10 \log \frac{60 \times 10^{-6}}{1.0125 \times 10^{-12}}$

$= 10 \log 59.26 \times 10^6 = 10(7.7727) = 77.7$ dB

34. $\text{dB} = 20 \log \frac{1.44}{1.73}$ or $\text{dB loss} = 20 \log \frac{1.73}{1.44}$

$= 20(-0.079\ 68)$ $= 20(0.079\ 68)$

$\therefore$ insertion loss = 1.59 dB   loss = 1.59 dB

36. Line attenuation $= 10 \log \frac{40 \times 10^{-6}}{10 \times 10^{-3}} = 10 \log 4 \times 10^{-3}$

$= -23.979$ dB

$\therefore$ circuit must add 23.979 dB From Prob. 35

$23.979 = 20 \log \frac{V_2}{V_1}$ $\therefore \frac{V_2}{V_1} = 10^{1.1989} = \underline{15.8}$

38. Number of Nepers $= \log_e \frac{I_1}{I_2} = 2.302\ 59 \log_{10} \frac{I_1}{I_2}$

No. of dB $= 20 \log \frac{I_1}{I_2}$

$$\frac{\text{No. of Nepers}}{\text{No. of dB}} = \frac{2.302\ 59 \cancel{\log \frac{I_1}{I_2}}}{20 \cancel{\log \frac{I_1}{I_2}}} = \frac{2.302\ 59}{20} = 0.115\ 129$$

$\therefore$ the no. of Nepers = 0.115 13 $\times$ the no. of dB

1 dB = 0.115 Nepers or 1 Neper $= (0.115)^{-1} = \underline{8.69 \text{ dB}}$

40. $\text{dB} = 10 \log \frac{8 \times 10^3}{500} = 10 \log 16 = \underline{12 \text{ dB}}$

42. $V \alpha (P)^{\frac{1}{2}}$ $\frac{V_2}{V_1} = \left(\frac{P_2}{P_1}\right)^{\frac{1}{2}} = \sqrt{2} = 1.414$

dB = 20 log 1.414 or dB = 10 log 2

$= 20(0.150\ 44)$ $= 10(0.301\ 03)$

$= \underline{3 \text{ dB}}$ $= \underline{3 \text{ dB}}$

PROBLEMS 35-3

2. $L = 0.621\ell\left[0.161 + 1.48 \log \frac{d}{r}\right] \times 10^{-3}$ H

$\ell = 30$ km, $d = 60$ cm, and $r = \frac{1}{2}(4.115) = 2.0575$ mm

$$L = 0.621 \times 30\left[0.161 + 1.48 \log \frac{600}{2.0575}\right] \times 10^{-3} \text{ H}$$

$L = 18.63(3.81) \times 10^{-3}$ $\therefore$ $\underline{L = 71 \text{ mH}}$

4. $\ell = 5$ km, No. 2 gauge copper $= 3.272$ mm $= r$, $d = 1$ m

Using formula in Prob. 2:

$$L = 0.621 \times 5\left(0.161 + 1.48 \log \frac{1000}{3.272}\right) \times 10^{-3} \text{ H}$$

$$\therefore L = 3.105(3.839\ 08) \times 10^{-3} = \underline{11.92 \text{ mH}}$$

$$C = \frac{0.0121\ell}{\log \frac{d}{r}} = \frac{0.0121 \times 5}{\log \frac{1000}{3.272}} = \underline{24.3 \text{ nF}}$$

6. $\ell = 22$ km, $d = 180$ mm, $r = 4.633$ mm

$$C = \frac{0.0121\ell}{\log \frac{d}{r}} = \frac{0.121 \times 2.2}{\log 38.85} = 0.1675\ \mu\text{F}$$

$$\text{Capacitance/km} = \frac{0.1675 \times 10^{-6}}{22} \qquad = \underline{0.007\ 61\ \mu\text{F/km}}$$

8. $C = 0.26\ \mu$F, $k = 4.3$, $d_2 = 8.252$ mm

$d_1 = (12.5)(2) + 8.252 = 33.252$ mm, $\ell = ?$

$$C = \frac{0.0241\ k\ell}{\log \frac{d_1}{d_2}}\ \mu\text{F} \qquad \therefore \ell = \frac{C \log \frac{d_1}{d_2}}{0.0241\ k} \text{ km}$$

Substitute values and solve for $\underline{\ell = 1.52 \text{ km}}$

10. $0.15 = \dfrac{0.0241 \times 4.3 \times \ell}{\log \dfrac{25 + d_2}{d_2}}$ (from Prob. 7)

$d_2 = 6.398$ mm, No. 2 copper wire $= 6.543$ mm $\therefore C = \dfrac{0.0241 \times 4.3 \times 5}{\log \dfrac{31.543}{6.543}}\ \mu$F

$\therefore C = 0.758\ \mu$F or $\underline{C = 758 \text{ nF}}$

12. $i = I_0 e^{-k\ell}$, $I_0 = 1.414i$ or $i = 0.707I_0$

$$\therefore\ 0.707I_0 = I_0 e^{-0.012\ell}$$

$$0.707\not{I}_0 = \not{I}_0 \frac{1}{e^{0.012\ell}} \qquad \therefore\ \log\ (0.707)^{-1} = 0.012\ell\ \log\ e$$

$$\log\ 1.414 = 0.005\ 21\ell$$

$$0.150\ 45 = 0.005\ 21\ell \qquad \therefore\ \underline{\ell = 28.9\ \text{km}}$$

14. $L = 9.21 \times 10^{-9} \log \frac{d}{r}$ H/cm $\qquad d = 140$ mm, $r = \frac{1.63}{2} = 0.815$ mm

$$L = 9.21 \times 10^{-9} \log \frac{140}{0.815} = 0.020\ 58\ \mu\text{H/cm}$$

For 250 m, $L = 250 \times 0.020\ 58 \times 100 = \underline{515\ \mu\text{H}}$

$$C = \left[9.21 \times 10^{-9} \times (3 \times 10^8 \times 10^2)^2 \log \frac{140}{0.815}\right]^{-1}\ \text{pF/cm}$$

$C = 0.053\ 98$ pF/cm. After 250 m, $\underline{C = 1.35\ \text{nF}}$

PROBLEMS 35-4

2. $Z_o = 276 \log \frac{d}{r} \qquad Z_o = 500\ \Omega$, $r = 0.645$ mm, $d = ?$

$$\frac{500}{276} = \log \frac{d}{0.645} \qquad \therefore\ \frac{d}{0.645} = 10^{\frac{500}{276}} \text{ and } \underline{d = 41.8\ \text{mm}}$$

Logical spacing would be 42 mm (4.2 cm)

4. $600 = 276 \log \frac{2d}{2.588} \qquad \therefore\ 14.925 = \frac{2d}{2.588}$ and $\underline{d = 193\ \text{mm}}$

6. No answer provided

8. $d = s + 2r$

$$Z_o = 276 \log \frac{s + 2r}{r} \qquad \therefore\ \log \frac{s + 2r}{r} = \frac{Z_o}{276} \qquad \therefore\ \frac{s + 2r}{r} = 10^{\frac{Z_o}{276}}$$

$$s = r10^{\frac{Z_o}{276}} - 2r \qquad \therefore\ r = s\left(10^{\frac{Z_o}{276}} - 2\right)^{-1}$$

$$\therefore\ r = \frac{40\ \text{mm}}{10^{\frac{200}{276}} - 2} = \frac{40}{12.217 - 2} = 3.92\ \text{mm} \qquad \therefore\ \underline{\text{tubing diameter} = 7.84\ \text{mm}}$$

10. $Z_o = 138 \log \frac{d_1}{d_2}\ \Omega$

$75 = 138 \log \frac{12}{d_2}$

$3.4953 = \frac{12}{d_2}$

$\underline{d_2 = 3.43 \text{ mm}}$

12. (a) Total loss = $0.2 \times 30 = 6$ dB

(b) $\% \text{ Eff} = \frac{P_{out}}{P_{in}} \times 100$

$10 \log \frac{P_{out}}{P_{in}} = -6 \text{ dB}$

$\frac{P_{out}}{P_{in}} = 10^{-0.6}$

$\therefore \% \text{ Eff} = \frac{100}{10^{0.6}} = \underline{25.12\%}$

14. (a) $\alpha = \frac{0.0157R_{ac}}{\log \frac{d}{r}}$ dB/m where $R_{ac} = \Omega\ ac$ for 1 m wire NOT line;

line = $R_{dc} \times 49$

$d = 30$ cm, $r = 1.632$ mm,

$R_{dc} = \Omega/\text{km} = 2.061\ \Omega/\text{km}$

$$\text{loss} = \frac{0.0157 \times 49 \times 2.061 \times 10^{-3} \times 460 \times 2}{\log \frac{300}{1.632}} \text{ dB}$$

$\underline{\text{loss} = 0.644 \text{ dB}}$

(b) $0.644 = 10 \log \frac{1000}{P_2}$ $\therefore P_2 = 862.15$ W Note: working step: divide by 10 ·

$\therefore\ 0.0644 = \log \frac{1000}{P_2}$

$\eta = \frac{862}{1000} \times 100 = \underline{86.2\%}$

16. (a) $$\alpha = \frac{38.33\sqrt{f}(d_1 + d_2)10^{-6}}{d_1 d_2 \log \frac{d_1}{d_2}} \text{ dB/m}$$

d_1 and d_2 in cm, f in MHz

$d_1 = 3.2$ cm, $d_2 = 0.8$ cm, $f = 27.8$ MHz, 400 m line

$$\alpha = \frac{38.33\sqrt{27.8}(3.2 + 0.8) \times 10^{-6}}{(3.2)(0.8)(\log 4)}$$

$\therefore \alpha = 0.5245 \times 10^{-3}$ dB/m

$\therefore$ for 400 m, $\underline{\text{loss} = 0.21 \text{ dB}}$

(b) Let $P_1 = 100$ W; then $0.21 = 10 \log \frac{100}{P_2}$ (divide by 10)

$\therefore 0.021 = \log \frac{100}{P_2}$ $\qquad \therefore P_2 = 95.3$ W

$\eta = \frac{95.3}{100} \times 100 = \underline{95.3\%}$

18. $R_{ac} = 8.33\sqrt{f}\left(\frac{1}{d_1} + \frac{1}{d_2}\right) \times 10^{-3}$ Ω/m

$f = 130$ MHz, $d_1 = 4.8$ cm, $d_2 = 0.38$ cm, line = 80 m

$R_{ac} = 0.2697$ Ω/m. For 80 m, $R_{ac} = \underline{21.6\ \Omega}$

20. Let $\frac{r\ell}{Z_o} = x$ $\qquad \therefore R_T = Z_o(e^x - 1)$ Ω. $\qquad Z_o = 90$ Ω, $\ell = 335$ m

r = RF resistance

= 0.33 Ω/m

$x = \frac{0.33 \times 335}{90} = 1.229$

$\therefore R_T = 90(e^{1.229} - 1)$

$= 217.6\ \Omega$

$\eta = \frac{100R_T}{R_T + Z_o} = \frac{100 \times 217.6}{307.6} = \underline{70.7\%}$

PROBLEMS 36-1

2. Sample solution: convert 001 010_2 to denary equivalent

$1 \times 2^3 = 8$

$0 \times 2^2 = 0$

$1 \times 2^1 = 2$

$0 \times 2^0 = 0$

$1010_2 = 10_{10}$

4. 11_{10} **6.** 39_{10} **8.** 49_{10} **10.** 61_{10}

PROBLEMS 36-2

2. Sample solution:

convert 12_{10} to binary

$$\begin{array}{r|rl} 2 & 12 & 0 \\ 2 & 6 & 0 \uparrow \\ 2 & 3 & 1 \\ & 1 & \end{array}$$

read up

$\therefore 12_{10} = 1100_2$

4. 23 = 10 111

6. 88 = 1 011 000

8. 126 = 1 111 110

10. 361 = 101 101 001

PROBLEMS 36-3

2. Sample solution:

convert 17_8 to denary

$(1 \times 8^1) + (7 \times 8^0) = 15_{10}$

4. $102_8 = 66_{10}$

6. $100_8 = 64_{10}$

8. $1035_8 = 541_{10}$

10. $22\ 453_8 = 9515_{10}$

PROBLEMS 36-4

2. Sample solution:

convert 37_{10} to octal

$$\begin{array}{r|rl} 8 & 37 & 5 \uparrow \\ & 4 & \end{array}$$

read up

$\therefore 37_{10} = 45_8$

4. $127 = 177_8$

6. $477 = 735_8$

8. $1062 = 2046_8$

10. $5000 = 11\ 610_8$

PROBLEMS 36-5

2. 30_4

4. 100_{16}

6. $202\ 221_3$

8. 5016_7

10. $10\ 142_5$

12. $163_7 = (7^2 \times 1) + (7 \times 6) + (7^0 \times 3) = 49 + 42 + 3 = \underline{94}$

14. $201_3 = (3^2 \times 2) + (3^0 \times 1) = 18 + 1 = \underline{19}$

16. $725_9 = (9^2 \times 7) + (9 \times 2) + (9^0 \times 5) = 567 + 18 + 5 = \underline{590}$

18. $2388_9 = (9^3 \times 2) + (9^2 \times 3) + (9 \times 8) + (9^0 \times 8) = 1458 + 243 + 72 + 8 = \underline{1781}$

20. $73006_8 = (8^4 \times 7) + (8^3 \times 3) + (8^0 \times 6) = (4096 \times 7) + (512 \times 3) + 6 = \underline{30\ 214}$

PROBLEMS 36-6

2. $\begin{array}{ccc} 2 & 7 & 7_8 \\ / & | & \backslash \\ 010 & 111 & 111 \end{array}$ $\quad 277_8 = 010\ 111\ 111_2$

4. 101 010 111

6. 110 100 101

8. 101 010 110 111

10. 001 000 000 000

12. 31_8

14. 3_8

16. 444_8

18. 117_8

20. 272_8

PROBLEMS 36-7

2. $\begin{array}{r} 100\ 101 \\ \underline{010\ 101} \\ \underline{\underline{111\ 010}} \end{array}$ 4. $\begin{array}{r} 0\ 110\ 110 \\ \underline{0\ 100\ 111} \\ \underline{\underline{1\ 011\ 101}} \end{array}$ 6. $\begin{array}{r} 101\ 111 \\ \underline{010\ 111} \\ \underline{\underline{1\ 000\ 110}} \end{array}$ 8. $\begin{array}{r} 110\ 010 \\ \underline{011\ 010} \\ \underline{\underline{1\ 001\ 100}} \end{array}$ 10. $\begin{array}{r} 100\ 101 \\ \underline{111\ 011} \\ \underline{\underline{1\ 100\ 000}} \end{array}$

PROBLEMS 36-8

2. $\begin{array}{rcr} 011\ 011 & = & 27 \\ \underline{-\ 010\ 111} & = & \underline{23} \\ \underline{\underline{100}} & = & \underline{\underline{4}} \end{array}$ 4. $\begin{array}{rcr} 110\ 111 & = & 55 \\ \underline{-\ 011\ 101} & = & \underline{29} \\ \underline{\underline{011\ 010}} & = & \underline{\underline{26}} \end{array}$ 6. $\begin{array}{rcr} 110\ 100 & = & 52 \\ \underline{-\ 101\ 111} & = & \underline{47} \\ \underline{\underline{000\ 101}} & = & \underline{\underline{5}} \end{array}$

8. $\begin{array}{rcr} 100\ 111 & = & 39 \\ \underline{-\ 100\ 011} & = & \underline{35} \\ \underline{\underline{100}} & = & \underline{\underline{4}} \end{array}$ 10. $\begin{array}{rcr} 100\ 110 & = & 38 \\ \underline{-\ 100\ 101} & = & \underline{37} \\ \underline{\underline{1}} & = & \underline{\underline{1}} \end{array}$

Problems 12, 14, 16, 18, and 20 are included beside problems 2, 4, 6, 8, and 10.

PROBLEMS 36-9

2. $\begin{array}{rr} & 101\ 101 \\ C = & \underline{101\ 101} \\ & 1\ 011\ 010 \\ & \underline{\llcorner\!\!\longrightarrow 1} \\ & \underline{\underline{011\ 011}} \end{array}$ 4. $\begin{array}{rr} & 001\ 101 \\ C = & \underline{111\ 001} \\ & 1\ 000\ 110 \\ & \underline{\llcorner\!\!\longrightarrow 1} \\ & \underline{\underline{000\ 111}} \end{array}$ 6. $\begin{array}{rr} & 101\ 011 \\ C = & \underline{110\ 101} \\ & 1\ 100\ 000 \\ & \underline{\llcorner\!\!\longrightarrow 1} \\ & \underline{\underline{100\ 001}} \end{array}$

8. $\begin{array}{rr} & 101\ 111 \\ C = & \underline{110\ 011} \\ & 1\ 100\ 010 \\ & \underline{\llcorner\!\!\longrightarrow 1} \\ & \underline{\underline{100\ 011}} \end{array}$ 10. $\begin{array}{rr} & 001\ 110 \\ C = & \underline{110\ 110} \\ & 1\ 000\ 100 \\ & \underline{\llcorner\!\!\longrightarrow 1} \\ & \underline{\underline{000\ 101}} \end{array}$ 12. $\begin{array}{r} 45 \\ \underline{18} \\ \underline{\underline{27}} \end{array}$

```
14.  13        16.  43        18.  47        20.  14
    - 6            - 10           - 12           - 9
    ---            ----           ----           ---
      7              33             35             5
```

PROBLEMS 36-10

```
2.      110 011        4.     010 111        6.      110 011
             11                   100                    110
      ---------        -------------        -----------------
        110 011        01 011 100                  1 100 110
       1 100 11                                   11 001 1
      ---------                             -----------------
      10 011 001                                100 110 010
```

```
8.      11 001 110         10.      111 001 111         12.  51
             1 101                        1 011               3
      ---------------           ------------------           ---
        11 001 110                  111 001 111              153
     1 100 111 0                 1 110 011 11
    11 001 110                111 001 111
    ---------------           ------------------
    101 001 110 110           1 001 111 100 101
```

```
14. 23        16. 51        18. 206        20. 463
     4             6             13              11
    --            ---           ----           ----
    92            306           2678           5093
```

PROBLEMS 36-11

```
2.          110                              4.            10 110
     101 | 011 110                               111 | 010 011 010
            10 1      Answer = 110                       1 11           Answer = 010 110
           -----                                        -------
             1 01                                        101 0
             1 01                                         11 1
           -----                                        -------
               00                                           111
                                                            111
                                                           ----
                                                             00
```

```
6.           0 000 101      8.             000 010 010     10.           100 101
     1111 | 1 001 011           010 010 | 101 000 100           10 111 | 1 101 010 011
              111 1                       100 10                           1 011 1
            ---------                    ------------                    -----------
              001 111                       100 10                         01 110 0
                1 111                       100 10                          1 011 1
            ---------                    ------------                     ----------
                   00                           00                             10 111
                                                                               10 111
                                                                               ------
                                                                                   00
```

Answer = 101_2

Answer = $10\ 010_2$

Answer = $100\ 101_2$

```
12.   6        14.   5        16.    5        18.    18        20.    37
    5|30           4|20          17|75           18|324           23|851
```

PROBLEMS 36-12

2. Sample solution: $\frac{4}{5} = 0.8_{10}$

0.8	→ 0.6	→ 0.2	→ 0.4	→ 0.8	→ 0.6	→ 0.2
× 2	× 2	× 2	× 2	× 2	× 2	× 2
1.6	1.2	0.4	0.8	1.6	1.2	0.4 → etc.!
↓	↓	↓	↓	↓	↓	↓
0.1	1	0	0	1	1	0_2

$\therefore 0.8_{10} = 0.110\ 011\ 0_2 = 1 \times 2^{-1} + 1 \times 2^{-2} + 1 \times 2^{-5} + 1 \times 2^{-6}$

$= 0.796\ 875_{10}$, which is 99.61% of 0.8

Answer is within ±1%; $0.110\ 011\ 0_2$ is -0.39% from 0.8_{10}

Show that 99% of $0.8_{10} = 0.110\ 010\ 101_2 = 0.791\ 99_{10}$

The following answers for Problems 4, 6, 8, and 10 are solved using the same method as for Prob. 2. Recall that 0.2_{10} is $0.001\ 100\ 110\ 011_2$.

4. $\frac{3}{5} = 0.6_{10} = 0.100\ 110\ 011_2 = 0.599\ 61_{10}$ (99.94% accurate)

6. $0.5_{10} = 0.100_2 = 0.5_{10}$ (100% accurate)

8. $\frac{9}{16} = 0.5625_{10} = 0.1001_2 = 0.5625_{10}$ (100% accurate)

10. $\frac{5}{11} = 0.454\ 545_{10} = 0.011\ 101_2 = 0.454\ 102_{10}$ (99.9% accurate)

PROBLEMS 36-13

Sample solution:

2. $\frac{11}{32} + \frac{4}{9} = 0.010\ 11_2 + 0.011\ 100\ 011_2 = 0.110\ 010\ 011_2$ (99.86% accurate)

$= \frac{99 + 128}{288} = 0.788194_{10}$ Convert this answer to binary using successive multiplication:

0.788 194	→ 0.576 388	→ 0.152 776	→ 0.305 552	→ 0.611 104	→ 0.222 208 →
× 2	× 2	× 2	× 2	× 2	× 2
1.576 388	1.152 776	0.305 552	0.611 104	1.222 208	0.444 416
↓	↓	↓	↓	↓	↓
0.1	1	0	0	1	0

0.444 416	→ 0.888 832	→ 0.777 664	→ 0.555 328	→ 0.110 656
× 2	× 2	× 2	× 2	× 2
0.888 832	1.777 664	1.555 328	1.110 656	0.221 312 →
↓	↓			
0	1			

Problem 36-13 Number 2, continued.

$\therefore 0.788\ 194_{10} = 0.110\ 010\ 011\ 100_2$ for 99.86% accuracy

Reconverting the binary number to denary we write:

$0.010 = 1 \times 2^{-2}$ $\quad\therefore 0.110\ 010\ 011\ 100_2$ is a denary number composed of:

$1 \times 2^{-1} + 1 \times 2^{-2} + 1 \times 2^{-5} + 1 \times 2^{-8} + 1 \times 2^{-9} + 1 \times 2^{-10} = \underline{0.788\ 086_{10}}$

Since accuracy to ±1% is all that is required, the remaining problems have solutions to within ±1% using the method shown.

4. $\dfrac{11}{132} + \dfrac{17}{24} = \dfrac{264 + 2244}{3168} = 0.791\ 667_{10} = 0.110\ 010\ 101\ 1_2$ (99.92% accurate)

6. $\dfrac{17}{32} - \dfrac{21}{64} - \dfrac{13}{64} = 0.203\ 125_{10} = 0.001\ 101_2$ (100% accurate)

   or $\dfrac{1}{8} + \dfrac{1}{16} + \dfrac{1}{64} = 0.001\ 101_2$ (100% accurate)

8. $\dfrac{59}{256} - \dfrac{987}{1024} = \dfrac{-751}{1024} = -0.733\ 398\ 44_{10} = -0.101\ 110\ 111_2$ (100% accurate)

10. $37.0875 - 27.125 = 9.9625_{10}$

    $100\ 101.000\ 1 - 11\ 011.001 = 1001.111\ 101_2$ (99.91% accurate)

12. $1.010\ 101_2 - 0.625_{10} = 1.328\ 125 - 0.625 = 0.703\ 125_{10}$

    $1.010\ 101 - 0.101_2 = 0.101\ 101_2$ (100% accurate)

    Complement $0.101\ 000 = 1.010\ 111$

    $$\begin{array}{r} \therefore\quad 1.010\ 101 \\ +\ 1.010\ 111 \\ \hline 10.101\ 100 \\ \llcorner\!\!\longrightarrow 1 \\ \hline 0.101\ 101_2 \end{array}$$

    Answer $= 0.101\ 101_2 \equiv 0.703\ 125_{10}$

    (100% accurate; see Prob. 7)

14. From Prob. 9: $101.0110 - 101.1111$ Answer is negative,

    $\therefore$ complement $101.0110 = 010.1001$

    $$\begin{array}{r} 101.1111 \\ -\ 010.1001 \\ \hline 1000.1000 \\ \llcorner\!\!\longrightarrow 1 \\ \hline 0.1001 \end{array}$$

    $\therefore$ Answer $= -0.1001_2 = -0.5625_{10}$

    (As in Prob. 9: 100% accurate)

PROBLEMS 36-14

2.
$$\begin{array}{r} 111 \\ \times\ 0.101 \\ \hline 111 \\ 111\ \ \\ \hline 100.011_2 \end{array} = \begin{array}{r} 7 \\ \times\ 0.625 \\ \hline 4.375_{10} \end{array}$$

4. $1_{10} = 1_2$

6. $0.195\ 312\ 5_{10} = 0.001\ 100\ 1_2$

8. $6.125_{10} = 110.001_2$

10. $0.351\ 562\ 5_{10} = 0.010\ 110\ 1_2$

PROBLEMS 36-15

2. Sample solution: $101.01_2 \div 0.011_2$ Move binary point:

$\therefore\ 101\ 010_2 \div 11_2$

$$\begin{array}{r} 1110_2 \\ 11_2 \overline{)101010} \\ 11 \\ \hline 100 \\ 11 \\ \hline 11 \\ 11 \\ \hline 00 \end{array}$$

$\therefore$ Answer $= 1110_2 = 14_{10}$ (100% accurate)

All other problems are solved using the same method as shown in this problem.

	Binary	Denary	Accuracy
2.	$1\ 110_2$	14_{10}	100%
4.	$1000.100\ 01_2$	$8.531\ 25_{10}$	99.98%
6.	$101\ 110.010\ 11_2$	46.345_{10}	100%
8.	$0.001\ 001\ 11_2$	0.1523_{10}	99.5%
10.	$10\ 000.1_2 + (1 \times 2^{-6})$	$16.516\ 13_{10}$	99.9% (100%)

Note: Except for Prob. 8 the accuracy can be stated to be 100% using five decimal places.

PROBLEMS 36-16

2. $110.\ 001\ 100_2$

$6\ .\ 1\ \ 4_8$

$= 6.14_8$

4. $1\ 011.100_2$

$1\ \ 3\ .\ 4_8$

$= 13.4_8$

6. 1001.011_2

$= 11.3_8$

8. $1.001\ 101_2$

$= 1.15_8$

10. $1000.101\ 101_2$

$= 10.55_8$

PROBLEMS 36-17

2. $396_{10} = 614_8 = 110\ 001\ 100_2$ binary triad

$= 110001100_2$ binary word

$= 0001\ 1000\ 1100_2$ binary tetrad

$= \ \ 1 \quad 8 \quad C_{16}$ hexadecimal

$\therefore 396_{10} = 18C_{16}$

Using the same method of solution:

4. $\underline{512_{10} = 200_{16}}$ (proof $200_{16} = 2 \times 16^2 + 0(16^1 + 16^0) = 512_{10}$)

6. $1C_{16} = 0001\ 1100_2$ binary tetrad

$= 011100_2$ binary word

$= 011\ 100_2$ binary triad $= 34_8$

*(See Prob. 5) $= (3 \times 8) + 4 = \underline{28_{10}}$

Using this method for Problems 8 and 10:

8. $\underline{53E_{16} = 1342_{10}}$ (proof $53E_{16} = 5 \times 16 + 3 \times 16 + 14$

$= 1280_{10} + 48_{10} + 14_{10} = 1342_{10}$)

10. $\underline{400_{16} = 1024_{10}}$ The proof is left as an exercise.

Emphasize: $A_{16} = 10_{10}$, $B_{16} = 11_{10}$, $C_{16} = 12_{10}$, etc.

*Note: **Reference for next problems: Tables 36-1, 2, and 3.**

PROBLEMS 36-18 Using Table 36-2 as outlined in text:*

2. $801_{16} = 2048_{10} + 0 + 1_{10} = \underline{2049_{10}}$

4. $AC1_{16} = 2560 + 192 + 1 = \underline{2753_{10}}$

6. $288_{10} = 288 - 256 = 32 = \underline{120_{16}}$

8. $560_{10} = 560 - 512 = 48 = \underline{230_{16}}$

10. $1056_{10} = 1056 - 1024 = 32 = \underline{420_{16}}$

PROBLEMS 36-19 Using Tables 36-2 and 36-3 as outlined in text: *

Convert to denary:

2. $0.B7C_{16}$

$$0.B = 0.687\ 5$$
$$0.07 = 0.027\ 343\ 750$$
$$0.00C = 0.002\ 929\ 688$$
$$\text{Sum} = \text{Answer} = 0.717\ 773\ 438_{10}$$

4. $0.56B_{16}$

$$0.5 = 0.312\ 50$$
$$0.06 = 0.023\ 437\ 50$$
$$0.00B = 0.002\ 685\ 547$$
$$\text{Sum} = \text{Answer} = 0.338\ 623\ 047_{10}$$

*Note: Not <u>all</u> procedural steps are shown;

*See examples 31 and 32 for Table 36-2 (text, p. 667) and examples 35, 36, and 37 for Table 36-3 (text, p. 670).

Convert to "hex":

6. 5002_{10} = 4096 ⟶ record 1_{16} hex pos 4

difference = 906

768 ⟶ record 3_{16} hex pos 3

difference = 138

128 ⟶ record 8_{16} hex pos 2

difference = 10 ⟶ record A_{16} hex pos 1

$\therefore 5002_{10} = 138A_{16}$

8. $1342_{10} = 1280_{10}$ ⟶ record 5_{16}

62_{10} ⟶ record 3_{16}

14_{10} ⟶ record E_{16}

$\therefore$ Answer $= 53E_{16}$

10. $15.796\ 142\ 578_{10}$

since $15_{10} = F_{16}$ first position to left is F.

0.796 142 578

0.750 ⟶ C_{16}

0.046 142 578

0.042 968 750 ⟶ $0B_{16}$

0.003 173 828 ⟶ $00D_{16}$

Combining the whole number and fractional portion:

$$15.796\ 142\ 578_{10} = F.CBD_{16}$$

PROBLEMS 37-1

2. $s\bar{\ell}$ 4. $\bar{s}\ell$ 6. $\bar{s}\bar{\ell}$ 8. $\overline{s + \ell}$ 10. $(s + \ell)(\bar{s}\bar{\ell})$

PROBLEMS 37-2

2.

a	b	c	$a + b$	$c(a + b)$
0	0	0	0	0
0	0	1	0	0
0	1	0	1	0
0	1	1	1	1
1	0	0	1	0
1	0	1	1	1
1	1	0	1	0
1	1	1	1	1

4.

a	b	c	$\bar{a}$	$\bar{b}$	$\bar{c}$	$a \cdot b \cdot c$	$a + b + c$	$\bar{a} + \bar{b} + \bar{c}$	$\overline{a \cdot b \cdot c}$
0	0	0	1	1	1	0	0	1	1
0	0	1	1	1	0	0	1	1	1
0	1	0	1	0	1	0	1	1	1
0	1	1	1	0	0	0	1	1	1
1	0	0	0	1	1	0	1	1	1
1	0	1	0	1	0	0	1	1	1
1	1	0	0	0	1	0	1	1	1
1	1	1	0	0	0	1	1	0	0

RHS = *LHS*

6.

a	b	c	$b + c$	ab	ac	$a(b + c)$	$ab + ac$
0	0	0	0	0	0	0	0
0	0	1	1	0	0	0	0
0	1	0	1	0	0	0	0
0	1	1	1	0	0	0	0
1	0	0	0	0	0	0	0
1	0	1	1	0	1	1	1
1	1	0	1	1	0	1	1
1	1	1	1	1	1	1	1

8.

a	b	ab	$a + ab$	a
0	0	0	0	0
0	1	0	0	0
1	0	0	1	1
1	1	1	1	1

10.

a	b	$\bar{a}$	$\bar{b}$	ab	$\overline{ab}$	$\bar{a} + \bar{b}$
0	0	1	1	0	1	1
0	1	1	0	0	1	1
1	0	0	1	0	1	1
1	1	0	0	1	0	0

12.

a	b	c	bc	$a + b$	$a + c$	$a + bc$	$(a + b)(a + c)$
0	0	0	0	0	0	0	0
0	0	1	0	0	1	0	0
0	1	0	0	1	0	0	0
0	1	1	1	1	1	1	1
1	0	0	0	1	1	1	1
1	0	1	0	1	1	1	1
1	1	0	0	1	1	1	1
1	1	1	1	1	1	1	1

14.

p	q	r	$\overline{q}$	$p + q$	$\overline{q} + r$	$q + 1$	$p\overline{q}$	qr	$(p + q)(\overline{q} + r)(q + 1)$	$p\overline{q} + qr$
0	0	0	1	0	1	1	0	0	0	0
0	0	1	1	0	1	1	0	0	0	0
0	1	0	0	1	0	1	0	0	0	0
0	1	1	0	1	1	1	0	1	1	1
1	0	0	1	1	1	1	1	0	1	1
1	0	1	1	1	1	1	1	0	1	1
1	1	0	0	1	0	1	0	0	0	0
1	1	1	0	1	1	1	0	1	1	1

16. $(a + c)(a + d)(b + c)(b + d) = ab + cd$

$LHS = RHS$

a	b	c	d	$a + c$	$a + d$	$b + c$	$b + d$	ab	cd	LHS	RHS
0	0	0	0	0	0	0	0	0	0	0	0
0	0	0	1	0	1	0	1	0	0	0	0
0	0	1	0	1	0	1	0	0	0	0	0
0	0	1	1	1	1	1	1	0	1	1	1
0	1	0	0	0	0	1	1	0	0	0	0
0	1	0	1	0	1	1	1	0	0	0	0
0	1	1	0	1	0	1	1	0	0	0	0
0	1	1	1	1	1	1	1	0	1	1	1
1	0	0	0	1	1	0	0	0	0	0	0
1	0	0	1	1	1	0	1	0	0	0	0
1	0	1	0	1	1	1	0	0	0	0	0
1	0	1	1	1	1	1	1	0	1	1	1
1	1	0	0	1	1	1	1	1	0	1	1
1	1	0	1	1	1	1	1	1	0	1	1
1	1	1	0	1	1	1	1	1	0	1	1
1	1	1	1	1	1	1	1	1	1	1	1

PROBLEMS 37-3

2. $(a + b)(\overline{a} + c)(b + c) = (\overline{a}b + ac)$

$LHS = RHS$

a	b	c	$\overline{a}$	$a + b$	$\overline{a} + c$	$b + c$	$\overline{a}b$	ac	LHS	RHS
0	0	0	1	0	1	0	0	0	0	0
0	0	1	1	0	1	1	0	0	0	0
0	1	0	1	1	1	1	1	0	1	1
0	1	1	1	1	1	1	1	0	1	1
1	0	0	0	1	0	0	0	0	0	0
1	0	1	0	1	1	1	0	1	1	1
1	1	0	0	1	0	1	0	0	0	0
1	1	1	0	1	1	1	0	1	1	1

4. $a(\bar{a} + b)(\bar{a} + b + c) = ab$

a	b	c	$\bar{a}$	$\bar{a} + b$	$\bar{a} + b + c$	$a(\bar{a} + b)(\bar{a} + b + c)$	ab
0	0	0	1	1	1	0	0
0	0	1	1	1	1	0	0
0	1	0	1	1	1	0	0
0	1	1	1	1	1	0	0
1	0	0	0	0	0	0	0
1	0	1	0	0	1	0	0
1	1	0	0	1	1	1	1
1	1	1	0	1	1	1	1

6. $\bar{q}t + \bar{q}\bar{t} + qt = \bar{q}(qt) + \bar{q}(\bar{q} \cdot \bar{t})$

$LHS = RHS?$

q	t	$\bar{q}$	$\bar{t}$	$\bar{q}t$	qt	$\bar{q}\bar{t}$	$\overline{qt}$	$\bar{q}(qt)$	$\bar{q}(\overline{qt})$	LHS	RHS
0	0	1	1	0	0	1	1	0	1	1	1
0	1	1	0	1	0	0	1	0	1	1	1
1	0	0	1	0	0	0	1	0	0	0	0
1	1	0	0	0	1	0	0	0	0	1	0

$LHS \neq RHS$ since $\bar{q}(qt) = 0$

8. $ABC + A\bar{B}C + AB\bar{C} + A\bar{B}\bar{C} + \bar{A}BC + \bar{A}\bar{B}C + \bar{A}B\bar{C} = A + B + C$

$LHS = RHS$ and let $X = ABC$

A	B	C	$\bar{A}$	$\bar{B}$	$\bar{C}$	$A\bar{B}C$	$A\bar{B}\bar{C}$	$\bar{A}BC$	$\bar{A}\bar{B}C$	$\bar{A}B\bar{C}$	$A + B + C$	LHS	RHS	X
0	0	0	1	1	1	0	0	0	0	0	0	0	0	0
0	0	1	1	1	0	0	0	0	1	0	1	1	1	0
0	1	0	1	0	1	0	0	0	0	1	1	1	1	0
0	1	1	1	0	0	0	0	1	0	0	1	1	1	0
1	0	0	0	1	1	0	0	0	0	0	1	1	1	0
1	0	1	0	1	0	1	0	0	0	0	1	1	1	0
1	1	0	0	0	1	0	1	0	0	0	1	1	1	0
1	1	1	0	0	0	0	0	0	0	0	1	1	1	1

10. $(\overline{ab + bc + ac}) = \bar{a} \cdot \bar{b} + \bar{b}\bar{c} + \bar{a} \cdot \bar{c}$ $RHS = LHS?$

a	b	c	$\bar{a}$	$\bar{b}$	$\bar{c}$	ab	bc	ac	$\bar{a}\bar{b}$	$\bar{b}\bar{c}$	$\bar{a}\bar{c}$	LHS	RHS
0	0	0	1	1	1	0	0	0	1	1	1	1	1
0	0	1	1	1	0	0	0	0	1	0	0	1	1
0	1	0	1	0	1	0	0	0	0	0	1	1	1
0	1	1	1	0	0	0	1	0	0	0	0	0	0
1	0	0	0	1	1	0	0	0	0	1	0	1	1
1	0	1	0	1	0	0	0	1	0	0	0	0	0
1	1	0	0	0	1	1	0	0	0	0	0	0	0
1	1	1	0	0	0	1	1	1	0	0	0	0	0

PROBLEMS 37-4

2.

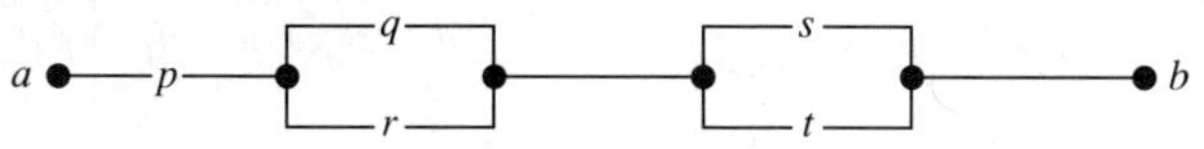

$Zab = \bar{p} + \bar{q}\bar{r} + \bar{s}\bar{t}$, $Yab = p(q + r)(s + t)$

4.

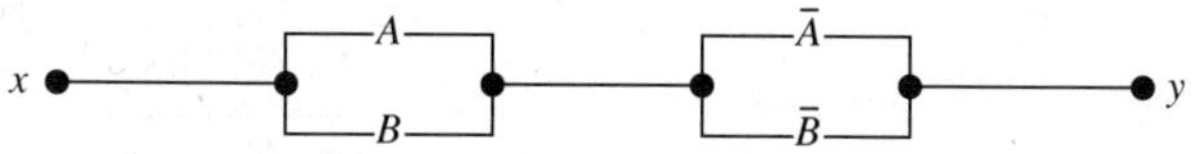

$Zxy = \bar{A}\bar{B} + AB$, $Yxy = A\bar{B} + \bar{A}B$

6.

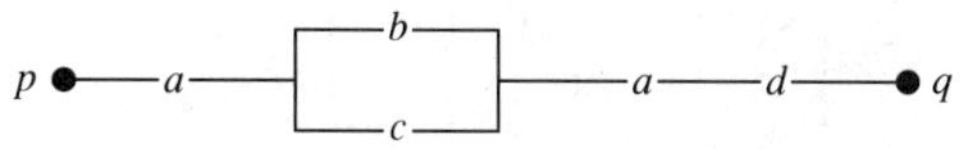

$Ypq = a(b + c)(ad)$

8.

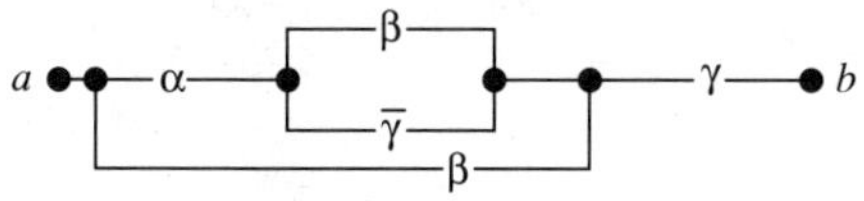

$Yab = [\alpha(\beta + \bar{\gamma}) + \beta]\gamma$

10.

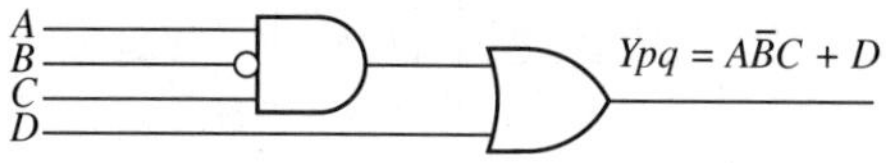

PROBLEMS 37-5

2.

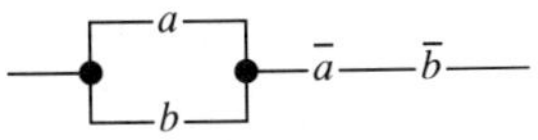

$(a + b)(\bar{a} \cdot \bar{b}) = \text{Function}$

$a\bar{a}\bar{b} + \bar{a}b\bar{b} = F$

But $a \cdot \bar{a} = 0$ and $b \cdot \bar{b} = 0$

$\therefore F = 0$ or open circuit

4.

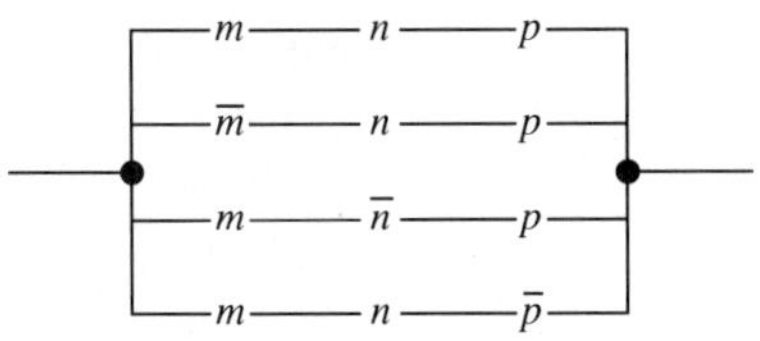

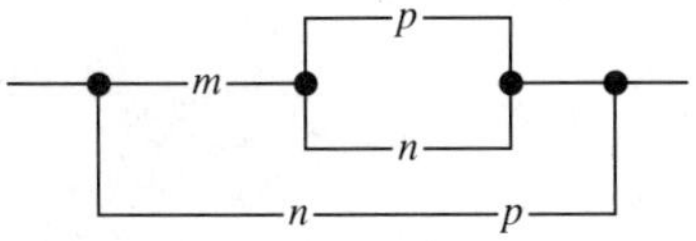

$F = mnp + \bar{m}np + m\bar{n}p + mn\bar{p}$

$= np\cancel{(m + \bar{m})} + m(\bar{n}p + n\bar{p})$

$\therefore F = np + m\cancel{(n + \bar{n})}(\bar{n} + \bar{p})\cancel{(p + \bar{p})}(n + p)$

$= np + (m\bar{n} + m\bar{p})(mn + mp)$

$= np + \cancel{mn\bar{n}} + m\bar{n}p + mn\bar{p} + \cancel{mp\bar{p}}$

$= np + m\bar{n}p + mn\bar{p}$

$= p(n + m\bar{n}) + mn\bar{p}$

$= p(n + m) + mn\bar{p}$

$= pn + pm + mn\bar{p}$

$= np + m(p + n\bar{p})$

$= np + m(p + n)$

$\therefore F = np + mp + mn$

6. $[(L + \overline{H} + \overline{K})(H) + K]HL = HL$

$LHS = RHS$

H	L	K	$\overline{H}$	$\overline{K}$	$L + \overline{H} + \overline{K}$	$(L + \overline{H} + \overline{K})(H)$	$(L + \overline{H} + \overline{K})H + K$	LHS	RHS
0	0	0	1	1	1	0	0	0	0
0	0	1	1	0	1	0	1	0	0
0	1	0	1	1	1	0	0	0	0
0	1	1	1	0	1	0	1	0	0
1	0	0	0	1	1	1	1	0	0
1	0	1	0	0	0	0	1	0	0
1	1	0	0	1	1	1	1	1	1
1	1	1	0	0	1	1	1	1	1

8. $F = pqr$

							a	b	c	
p	q	r	pqr	pq	qr	pr	$pq + qr + pr$	$pqr(a)$	$(pqr + pq)b$	$b + c$
0	0	0	0	0	0	0	0	0	0	0
0	0	1	0	0	0	0	0	0	0	0
0	1	0	0	0	0	0	0	0	0	0
0	1	1	0	0	1	0	1	0	0	0
1	0	0	0	0	0	0	0	0	0	0
1	0	1	0	0	0	1	1	0	0	0
1	1	0	0	1	0	0	1	0	0	0
1	1	1	1	1	1	1	1	1	1	1

$Y = pqr(pq + qr + pr) + pqr(pqr + pq)$

$= pqr + pqr + pqr$

$\underline{Y = pqr}$

10. I $= ab(cd + \bar{c}d + \bar{c}\bar{d})$ II $= bc(ad + a\bar{d})$

a	b	c	d	$\bar{c}$	$\bar{d}$	ab	bc	$\bar{c} + d$	cd	$\bar{c}d$	$\bar{c}\bar{d}$	$a\bar{d}$	ad	$ad + a\bar{d}$	$ab(\bar{c} + d)$	I	II	I + II
0	0	0	0	1	1	0	0	1	0	0	1	0	0	0	0	0	0	0
0	0	0	1	1	0	0	0	1	0	1	0	0	0	0	0	0	0	0
0	0	1	0	0	1	0	0	0	0	0	0	0	0	0	0	0	0	0
0	0	1	1	0	0	0	0	1	1	0	0	0	0	0	0	0	0	0
0	1	0	0	1	1	0	0	1	0	0	1	0	0	0	0	0	0	0
0	1	0	1	1	0	0	0	1	0	1	0	0	0	0	0	0	0	0
0	1	1	0	0	1	0	1	0	0	0	0	0	0	0	0	0	0	0
0	1	1	1	0	0	0	1	1	1	0	0	0	0	0	0	0	0	0
1	0	0	0	1	1	0	0	1	0	0	1	1	0	1	0	0	0	0
1	0	0	1	1	0	0	0	1	0	1	0	0	1	1	0	0	0	0
1	0	1	0	0	1	0	0	0	0	0	0	1	0	1	0	0	0	0
1	0	1	1	0	0	0	0	1	1	0	0	0	1	1	0	0	0	0
1	1	0	0	1	1	1	0	1	0	0	1	1	0	1	1	1	0	1
1	1	0	1	1	0	1	0	1	0	1	0	0	1	1	1	1	0	1
1	1	1	0	0	1	1	1	0	0	0	0	1	0	1	0	0	1	1
1	1	1	1	0	0	1	1	1	1	0	0	0	1	1	1	1	1	1

$\frac{o}{p}$ becomes I + II as shown

$o/p = ab(\bar{c} + d) + abc = ab(\bar{c} + c) + abd$

$o/p = ab$

Does this agree with answer to Prob. 5?

Solution by Boolean algebra shows that the function reduces to a single AND gate $a \cdot b$ since there is a path with or without c or d (c and d are "don't care" or redundant states).

PROBLEMS 37-6

2.

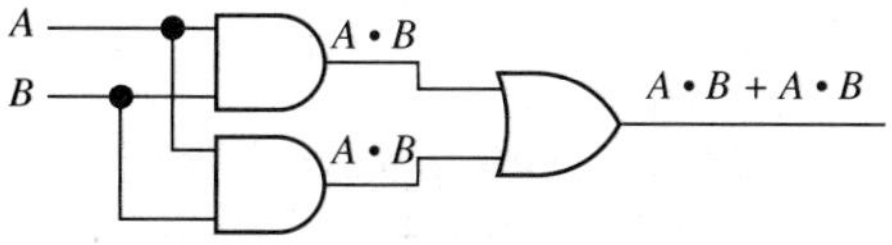

$A \cdot B + A \cdot B = A(B + B) = A \cdot B$

∴ replace with a single AND gate

4. $C_o = \bar{a}bc + a\bar{b}c + ab\bar{c} + abc$

$= c(\bar{a}b + a\bar{b}) + ab(\bar{c} + c)$ Ignore XOR at this time!

$= c(b + a) + ab$

$\therefore C_o = bc + ac + ab$

This eliminates three INVERTERS and one AND gate.

PROBLEMS 37-7

2.

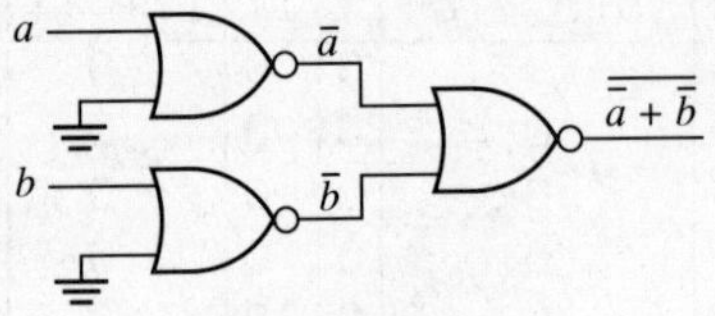

$\overline{\bar{a} + \bar{b}} = \overline{\overline{a \cdot b}} = ab$
single AND gate

4.

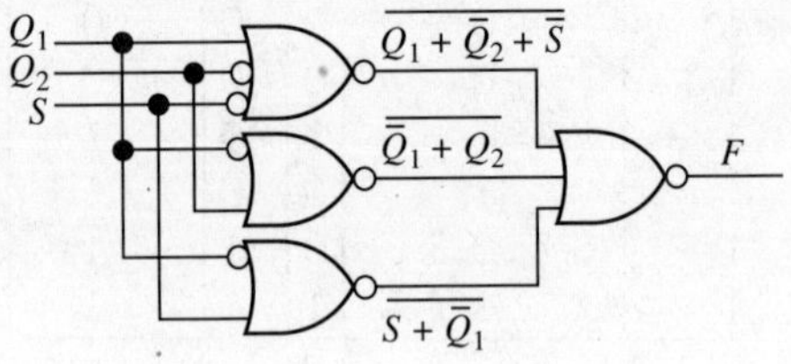

$$F = \overline{Q_1 + \bar{Q}_2 + \bar{S}} + \overline{\bar{Q}_1 + Q_2} + \overline{S + \bar{Q}_1}$$

$$= \bar{Q}_1 Q_2 S + Q_1 \bar{Q}_2 + \bar{S} Q_1$$

$$= \bar{Q}_1 (Q_2 S) + Q_1 (\bar{Q}_2 + \bar{S})$$

$$= \bar{Q}_1 (Q_2 S) + Q_1 (\overline{Q_2 S})$$

Which is of the form:

$$\bar{A}B + A\bar{B}$$

indicating an Exclusive OR.

$$F = \bar{Q}_1 (Q_2 S) + Q_1 (\overline{Q_2 S})$$

PROBLEMS 38-1

2. $A\bar{B} + \bar{A}B$

4. $\bar{A}\bar{B}\bar{C} + A\bar{B}C + A\bar{B}\bar{C} + ABC$

6.

	$\bar{A}$		A	
$\bar{B}$	1	1		
B		1	1	1
	$\bar{C}$	C	C	$\bar{C}$

8.

		$\bar{A}$		A		
$\bar{B}$		1		1	1	$\bar{D}$
$\bar{B}$			1			D
B			1			D
B		1	1	1	1	$\bar{D}$
		$\bar{C}$	C	C	$\bar{C}$	

PROBLEMS 38-2

2. By tautologies:

$$AB(\bar{C} + C) + BC(\bar{A} + A) = AB + AC$$

The term $\bar{A}C$ that is derived from the map is already grouped within the other two terms, and is considered REDUNDANT when performing the function.

	$\bar{A}$		A	
$\bar{B}$	1	1		
B		1	1	
	$\bar{C}$	C	C	$\bar{C}$

Three sets of pair groupings to show that: function = $\bar{A}\bar{B} + \bar{A}C + BC$

4.

	$\bar{A}$		A		
$\bar{B}$	1	1			$\bar{D}$
$\bar{B}$		1			D
B		1	1		D
B	1	1		1	$\bar{D}$
	$\bar{C}$	C	C	$\bar{C}$	

$F = \bar{A}\bar{D} + \bar{A}C + BCD + B\bar{C}\bar{D}$

By tautologies:

$$F = \bar{A}\bar{B}\bar{C}\bar{D} + \bar{A}\bar{B}C\bar{D} + \bar{A}\bar{B}CD + \bar{A}BCD + ABCD + \bar{A}B\bar{C}\bar{D} + \bar{A}BC\bar{D} + AB\bar{C}\bar{D}$$

$$= \bar{A}\bar{B}\bar{C}\bar{D} + \bar{A}\bar{B}C(\bar{D} + D) + BCD(\bar{A} + A) + \bar{A}BC\bar{D} + B\bar{C}\bar{D}(\bar{A} + A)$$

$$= \bar{A}\bar{B}\bar{C}\bar{D} + \bar{A}\bar{B}C + BCD + \bar{A}BC\bar{D} + B\bar{C}\bar{D}$$

$$= \bar{A}\bar{B}(\bar{C}\bar{D} + C) + BC(D + \bar{A}\bar{D}) + B\bar{C}\bar{D}$$

$$= \bar{A}\bar{B}(\bar{D} + C) + BC(D + \bar{A}) + B\bar{C}\bar{D}$$

$$= \bar{A}\bar{B}\bar{D} + \bar{A}\bar{B}C + BCD + \bar{A}BC + B\bar{C}\bar{D}$$

$$= \bar{A}\bar{B}\bar{D} + \bar{A}C(\bar{B} + B) + BCD + B\bar{C}\bar{D}$$

$$F = \bar{A}\bar{B}\bar{D} + \bar{A}C + BCD + B\bar{C}\bar{D}$$

Other groupings will provide:

$$F = \bar{A}\bar{B}\bar{D} + \bar{A}\bar{B}C + BCD + B\bar{C}\bar{D} + \bar{A}BCD$$

or $F = \bar{A}\bar{B}\bar{D} + \bar{A}\bar{B}C + BCD + B\bar{C}\bar{D}$

6. Karnaugh map of the function:

$$F = \bar{A}BC + A\bar{B}C + AB\bar{C} + A\bar{B}\bar{C} + \bar{A}\bar{B}C + \bar{A}B\bar{C} + ABC$$

6 8 3 4 5 2 7

	$\bar{A}$		A	
$\bar{C}$		1	1	1
C	1	1	1	1
	$\bar{B}$	B		$\bar{B}$

Horizontal pair-groupings:

Boxes 2 and 3 to produce $B\bar{C}$

Boxes 3 and 4 to produce $A\bar{C}$

Boxes 5 and 6 to produce $\bar{A}C$

Boxes 6 and 7 to produce BC

Boxes 7 and 8 to produce AC

Function becomes: $F = B\bar{C} + A\bar{C} + \bar{A}C + BC + AC$

Applying tautological simplification:

$F = \bar{A}C + B(C + \bar{C}) + A(C + \bar{C})$

$= \bar{A}C + B + A$, and since $A + \bar{A}C$ is tautological with $A + C$, the function becomes $\underline{A + B + C}$

Note: Prob. 6 solution showed that the use of pair-groups and tautological simplification on the resulting expression, $F = \bar{A}BC + A\bar{B}C + AB\bar{C} + A\bar{B}\bar{C} + \bar{A}\bar{B}C + \bar{A}B\bar{C} + ABC$ quickly reduces/simplifies to: $\underline{F = A + B + C}$

8. By tautologies:

Carry function of Fig. 37-20

$$F = ab\bar{c} + \bar{a}bc + a\bar{b}c + abc$$

$F = ab(c + \bar{c}) + \bar{a}bc + a\bar{b}c$

$= ab + \bar{a}bc + a\bar{b}c$

$= b(a + \bar{a}c) + a\bar{b}c$

$= ba + bc + a\bar{b}c$

$= ba + c(b + a\bar{b})$

$= ba + bc + ac$

Alternative grouping:

$F = bc(a + \bar{a}) + ab\bar{c} + a\bar{b}c$

$= bc + ab\bar{c} + a\bar{b}c$

$= b(c + a\bar{c}) + a\bar{b}c$

$= bc + ab + a\bar{b}c$

$= bc + a(b + \bar{b}c)$

$= bc + ab + ac$

Both routines/groupings produce the same result, and the carry function contains no inverted terms.

By Karnaugh map: $F = \underset{3}{a\bar{b}c} + \underset{6}{\bar{a}bc} + \underset{7}{abc} + \underset{8}{ab\bar{c}}$

	$\bar{a}$	$\bar{a}$	a	a
$\bar{b}$			1	
b		1	1	1
	$\bar{c}$	c	c	$\bar{c}$

Pair-groupings:

Horizontal-pairs:

Boxes 6 and 7 produce bc

Boxes 7 and 8 produce ab

Vertical-pairs:

Boxes 3 and 7 produce ac

Function becomes: $F = bc + ab + ac$

If the vertical pair-group is omitted, the simplified expression is:

$$F = bc + ab + a\bar{b}c$$

which reduces to: $F = ab + c(b + a\bar{b})$

$= ab + c(b + a)$

$\therefore F = ab + bc + ac$

The vertical pair-grouping removed the need to apply/use tautological simplification to reduce the function to: $F = ab + bc + ac$.

(Did you notice? Box 7 is grouped three times.)

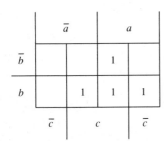